TAILS FROM THE TRACK

TAILS FROM THE TRACK

A Journey of Self-Discovery on the Backstretch

Janice Gannon

CANTER ON PRESS

Copyright © 2016 Janice Gannon
All rights reserved. This book or any portion thereof may not be reproduced or used in any manner whatsoever without the express written permission of the publisher except for the use of brief quotations in a book review.

Book layout and design by Ward Edwards (www.BookForge.ca)
Illustrations and cover design by Nola McConnan aoca aestc tws (www.merriweatherdesignstudio.com)

Canter On Press
www.CanterOnPress.com
info@CanterOnPress.com

Library and Archives Cataloging-in-Publication

Gannon, Janice, 1957-, author
       Tails from the track : a journey of self-discovery on the backstretch / Janice Gannon.

ISBN 978-0-9951890-0-3 (paperback)–ISBN 978-0-9951890-1-0 (hardback)

       1. Gannon, Janice, 1957- –Anecdotes. 2. Horse grooms–Ontario–Anecdotes. 3. Racetracks (Horse racing)–Ontario–Anecdotes. 4. Horse racing–Anecdotes. I. Title.

SF336.G36A3 2016   798.40092   C2016-903892-0

*To "the gang" - Sandy, Sue, Henrietta and Donny.
Your friendship supported me through my rough days and gave me
the courage to reinvent myself.*

Contents

Preface		xi
1	Horse Crazy	1
2	Swing Groom	7
3	Greenwood Spring Meet	17
4	The Little Bird of a Filly	29
5	Number 47	37
6	Not the Best	45
7	Aurora Meadows	51
8	Mr Annesley	63
9	Summer of Dispair	75
10	Undervalued	79
11	Scottish Music	85

12	No Sense of Humour	91
13	At the Gate	95
14	Mistakes	105
15	Fort Dreary	111
16	The Letter	121
17	Winter of Reflection	127
18	Working with Sandy at Greenwood	135
19	Working With Sandy, Summer	143
20	The Scamp	153
21	Lazy	157
22	Making Ends Meet	163
23	Doing What It Takes	171
24	Getting My Break	183
25	Red Millar's Cat	189
26	Galloping Lessons	195
27	Heading South	203
28	A Brand New Farm	209
29	Proving Myself	213
30	Laid Up and Laid Off	221

31	Never Right	229
32	Track Life	235
33	What a Trip	241
34	We, the Anonymous	247
Glossary		251
Appendix 1 - Format of Racing		265
Appendix 2 - Woodbine		269
Acknowledgements		271

Preface

This story is about my years at Woodbine (and Fort Erie and Greenwood) racetracks, from September 1977 to the late summer of 1986. During this period, Thoroughbred racing was experiencing a renaissance. Money was plentiful then and many champions emerged.

I have written this as a sort of you-are-there experience. It is a window into a time that is rapidly vanishing. Because I want you to feel what life was like on the backstretch, I have deliberately used the expressions we used, even when grammatically incorrect. Because of the amount of jargon, I have made an extensive glossary. The first time a word or expression appears which is likely unfamiliar to you, I have put any such word or expression in bold type that appears in the glossary.

This is my personal story and I make no apologies for information that is biased or wrong. Memory is not perfect. People can remember the same incident very differently, according to their point of view. This is my story and the incidents are as accurate as I recall.

Although there have been books written about the backstretch, they were mostly by newspaper men or else written by outsiders, people temporarily employed there. I have yet to read another account by a groom or exercise rider. It was and is a totally unique environment.

This is also about my own journey, of growing up and acquiring the skills I needed to be successful in life. The lessons learned were subsequently parlayed into a rewarding career teaching riding and a happier personal life.

I have used pseudonyms for about half of the people in my story. I worked for a number of men named Robert or John. For clarity, I have used different first names. Whenever I portrayed someone in an unflattering manner, I substituted another name. Otherwise, I used the real names of some of the trainers because they are historical people. For most of the backstretch personnel I used their real first name but not much else in the way of identifying information. Some people I have used their actual names with their permission.

Except for Briar's Boy and Treya, all the horses names were either their actual names or their common barn names.

Now, come with me into the world of racehorses.

Chapter One

Horse Crazy

It's September and I have just been fired for the first time. Fresh out of Humber College's Horsemanship Program just four months ago, I had begun working on a well known horse breeding farm. It was a dream job except for my cantankerous elderly boss for whom I was to prove equally as stubborn. By summer's end, we came to a mutual parting of ways. Since I was now supporting myself, I needed a job, fast and preferably one with horses.

I was told the racetrack is always hiring and I decide to try the racetrack option. What I do not know is that by September, most of the trainers are downsizing their operations, as horses break down and are moved out. Along with the warmer weather, many jobs are disappearing for the season.

The backstretch is a world unto itself and the raw public is excluded. Everyone needs a badge to get past the gate security. Nonetheless, I am sitting in the crude construction trailer which serves as a personnel office. It is located just inside the chain link fence that surrounds the backstretch at Woodbine race track. The interior of the trailer is chilly and nearly bare. Against the end wall a steel desk huddles with chairs stiffly arranged on either side. Behind the desk is a calendar, under the calendar sits a filing cabinet.

Beside them, a small heater throws out a weak heat. This is the office of Jimmy, a semi-invalid whose job is to sort out the competent from the criminal and the riff raff and to match available jobs on the back stretch to the applicants. I feel like an alien in the starkly decorated office.

"Do you have any experience?" Jimmy inquires gently.

I reel off my short resume. Graduate of Humber College's Equine Program, rubbed **yearlings** for Windfields Farms, riding and breaking yearlings, a stint on a breeding farm.

"I might have something for you. Beasley Farms needs a swing groom. It is not full time but it may be just the ticket for you."

Jimmy leads me outside of the trailer where I climb onto the seat of his golf cart. The sun has not yet peeked over the horizon but I can just see the horses galloping on the training track, black silhouettes against a pale pink-and-baby blue sky. The entire back stretch is bubbling with activity as I am quietly whisked down to Barn 7.

* * *

I cannot remember a time when I was not passionate about horses. It was as if I popped out of the womb worshipping them. While other little girls played house, my daydreams were all about horses. It was a time when television was dominated by 'westerns' with my favourite cowboy being Roy Rogers. My teen age sister Dee read horse books aloud to me. My sister Anna knew the best way to get even with me was to torture my stuffed toy horse named Dark Joel.

When asked what they wanted to be when they grew up, the girls of my generation primly answered, "Teacher" or "Secretary" or "Mother." My teachers were clearly puzzled by my reply of "Cowboy". Okay, there was this little matter of the masculine persuasion. Clearly, God had made a mistake when handing out the gender allotment. After all, it was my hero Roy Rogers who rode the stunning palomino Trigger and caught the bad guys while his wife Dale Evans rode in after all the action was finished.

Eventually, I was persuaded that being female was something I was going to have to live with but I stayed determined to own a horse. I made deals with God (I felt that he owed me one) and pestered my parents. I read somewhere that if I believed I would receive something badly enough, God would come through.

Gradually, I learned that my dad was only a blue collar worker trying to feed five children and besides, we lived in the city. I let go of my dream.

In my real world, I was witness to some terrible fights between my parents. My sister Anna and I were sometimes used as bargaining tools to keep my mother from walking out on us. The fights ended abruptly with The Fire.

The Fire happened on Boxing Night, when my parents had uncharacteristically gone out to a movie. My parents rented out rooms in our too big house to pay the mortgage. The tenant in the attic often babysat Anna and me. This night, however, my parents left Dee in charge. When Dee came up to bed, she discovered smoke seeping from the door to the attic. She frantically roused the rest of the household but the smoke barricaded her from the attic apartment. Anna and I were sent out onto the front porch.

Smoke silently curled down the stairs in gray plumes as I watched through the front door's glass. Dee then sent us next door to neighbours for shelter. As we picked our way through the snow, in our pajamas and bare feet, I saw the sparks cascading from the back of the house. My safe world would never be the same.

When we were summoned home next morning, Mom was crying as Dad gently broke the news. Our tenant, who was both my friend and sometime babysitter, had died in the blaze.

There were no teams of social workers parachuted in to help us cope with the trauma, despite The Fire making the front page news. When Anna and I returned to school, we were still deeply in shock. We withdrew from our boisterous classmates, nursing our pain. Our fragile world had cracked apart like an egg. Our self absorbed parents never noticed how our classmates now reacted to our changed behaviour. Suddenly, we were vulnerable targets and our classmates bullied us relentlessly.

Animals kept me sane. Our pets provided the love that was in short supply elsewhere in my life. There was always a cat and a dog and sometimes a whole parade of other animals that lived with us, including an unbroken pony. I depended on their their unstinting devotion to me. Slowly, I healed.

Two personality traits emerged as a result of my classmates' ostracism. I had developed an inner resilience beyond my years. I was also determined to forge my own destiny and would not be dissuaded by the opinions of my peers.

I was eight when Dad surprised the family by buying a pony. While I would have to share her with Anna, this did not inhibit my joy. The arrival of the pony convinced me that my steadfast belief in getting a horse had created the real deal. Having the pony was a slice of rare magic in my young life.

The real pony was very different from the story books. Star was a youngster who had never been ridden. My parents knew next to nothing about horses and we had no one to advise us. We kept the pony at my grandmother's farm in the summer and found a cheap stable just a couple of miles from my house. After a year of gentling the pony, my parents bought a western saddle and strapped it on. Then they put each of us in the saddle for a ride. When the pony laid down with Anna on her, my parents beat the pony for her misbehaviour.

I was deeply shocked. I mulled the incident over for days. Finally, I told my parents we were going to start over and slowly introduce Star to bridle and saddle. Star never misbehaved in this way again.

We made all the ignorant mistakes of newbies. We spoiled Star until she bit us, I had my share of falls. Several years later I **foundered** her, letting her drink her fill of icy water when she was hot because I was in a hurry.

I rode in a western saddle until I outgrew it, then rode bareback. By age twelve, Anna and I had begun to save for an English saddle, hoping to learn jumping. For two years, we collected pop bottles and banked any other cash that came our way. Before we had enough money for a new saddle, I was fourteen and had outgrown the pony so we sold her. Delivering Star to her new home was one of the saddest days of my life.

My frustration with Star was never having anybody knowledgeable to teach me. The pony was smart and willing but I had little idea how to better train her. I had not learned much about riding but I had learned some about horses.

Our saddle money sat in the bank until the following summer when a new riding stable opened nearby. They ran an ad in the local paper for riding lessons. The money saved for a saddle now paid for ten lessons each for Anna and me. After the first set of lessons ran out, I knew I had to find a way to keep riding. I calculated I could afford to ride if I spent my clothing allowance on the lessons. Giving up fashionable clothes was not a hard choice for me.

My first instructor was a European ex-cavalry man with a thick accent and little talent for teaching but who managed to generate wild enthusiasm among the young riders. He was replaced a year later by a much younger man who was the complete opposite. My second instructor was a polished equitation rider who was horrified by the wild excesses of our riding group. We soon shaped up under his tutelage.

Still, a lesson once a week was a long way from a career with horses. Then one night, my Dad called me urgently to the TV. The news was profiling a course in horsemanship at Humber College. Watching the television with growing excitement, I knew this was my ticket to my career.

That summer, using my plan to attend Humber as leverage, I landed a job at Windfields Farm. This was the premier thoroughbred breeding farm in Canada located just a few miles away. By the end of summer, I could adequately muck a stall, turn out a horse safely and handle most difficult horses from the ground. I rubbed yearlings for the sales and accompanied them to the September sale held on the far side of Woodbine racetrack. We were billeted in trailers and ate in the track kitchen. Naive to the racing world, I was terrified of the scruffy racetrack people I saw there. Many of them were unshaven, missing teeth, wearing ripped and unwashed clothes. There were few girls among them. I stayed close to my co-workers as we walked through the backstretch to the kitchen for each meal.

The trailers that we stayed in overlooked the training track at Woodbine. Every morning upon wakening, I could see the silhouettes of horses galloping against a pale sky. As the horses came by, I could hear their staccato beat – thud-ud, thud-ud, thud-ud. It was as if somehow through the night, magical carousel horses had come to life. This was better than anything I could have dreamed of. The early morning gallops were in sharp contrast to the seedy backstretch area.

I had no idea that four years later, I would be back.

Chapter Two

SWING GROOM

Now I am back at Woodbine. I stand outside the shedrow while Jimmy asks for Tommy, the trainer. Tommy proves to be a perpetually sadder, older man, just putting in his time before he can retire. Tommy asks me the same questions as Jimmy did and I give the same answers. He surprises me by telling me he already has two girls from Humber's equine program working for him. I know both Sam and Linda from school. Tommy explains that there are four full time employees plus a swing groom. The swing groom works part time only on their days off. With no other options available, I accept the job. Days off are a new concept at the track with only a minority of the stables offering their help a day off.

This barn has one other bizarre feature. While Warren Beasley owns all the thoroughbreds in the stable, the training is split between two trainers. Whether this is because the owner has chosen to spur competition between the trainers in this manner or to charitably show loyalty to his father's trainer, I am not sure. Tommy has been with the outfit for years but Dean is the better trainer. The twelve horses are evenly divided between them. Duck works for Tommy and Linda for Dean. Sam is literally and figuratively between them, **rubbing** two horses for each trainer.

Rounding out the complement of people are Jack, the handy man and Jimmy, the contract jockey. Jimmy gallops some of our horses, works all of them and then rides them in the afternoon (in the races). I fortunately have few dealings with him because neither of us care much for the other.

Jack feeds the horses at 5 am, tops up the water at night, holds horses for their morning baths and babysits the **hot walking machine**. The machine is like a carousel and has space for four horses to cool out at the same time. Jack is loud, somewhat obnoxious and truly incompetent at handling horses. Only his willingness to live at the barn and feed the animals guarantees his job. There is one blessing to this job. Both Duck and Linda are top notch grooms. They want their horses expertly looked after and more than adequately train me to care of them. I spend a day shadowing each of these co-workers before I am turned loose.

Duck and Linda represent the two types of grooms found at the track. Duck (the only name I know her by) has been working on the backstretch for years. She is older than I, as tough as old leather and speaks her mind freely with plenty of swear words mixed in. Her clothes look like they come from Salvation Army and haven't seen a washing machine in a while, with white poultice wiped down her jeans. I am not sure if she finished public school.

Linda typifies the newer grooms on the track. Her good looks and lively personality are popular with both men and women. She is smart and college educated (being a Humber graduate). Linda dresses neatly, even applying some makeup. Single, outgoing and fun-loving, she is frequently invited to parties. I feel more comfortable around her. The commonality of both grooms is that they are utterly devoted to their horses.

There can hardly be two more different trainers than Dean and Tommy. Dean can send an older, patched-back-together cripple (broken down horses are routinely called 'cripples') over to race and the girls confidently place their bets to **win**. With Tommy, winning seems more by luck than design and the girls are more apt to keep their cash in their pockets. When Sam's two year old colt, trained by Tommy, wins, it is completely unexpected. As Sam is doing up the colt after the race, Tommy gazes sadly at the young horse, trying to determine what he did differently this time to win.

The rivalry between the two trainers creates a poisonous atmosphere. While they are coolly polite on the surface, Linda and Duck have no such conventions. The trainers act as puppeteers, pulling the strings of their respective grooms. The grooms fight noisily like angry cats over everything; which horses go out first, which horses are to be shod by the blacksmith today. The spats are loud, vicious and frequent. I also notice Linda usually wins.

Sam is smart enough to simply keep her head down and her mouth shut. Sam never expresses her own wishes. Having just been promoted to groom from swing groom, she knows enough not to get pulled into the politics of the barn.

Woodbine track is a huge complex, perhaps two square miles in size.[1] The racing surfaces consist of the mile long main track, the inner turf course and the Marshall turf course. The backstretch has another mile long training track, an eighth mile oval sand ring, and a back field with a narrow path ploughed around it. There are fourteen horse barns, a drug test barn, a track kitchen, a racing office, tack stores, vet offices and maintenance equipment storage. Green lawns surround the buildings, adding to the spacious feeling of the place.

The barns are paired (as in 1 and 1A) and joined by a series of 6'x8' rooms so that the two barns form an H pattern when viewed from above. These rooms are used as offices, tack rooms but most commonly as living quarters for the male grooms and hotwalkers that populate the backstretch. These tack rooms open directly onto the outdoors. Girls are forbidden on the backstretch after 6 pm but have a newly built dormitory of their own. Its location, just inside the front gate, allows the security guards to keep an eye on them. Women's liberation has not fully infiltrated the backstretch.

At Woodbine, each barn consists of 48 stalls. The stalls are arranged back to back, with a dirt aisle, called the **shedrow**, surrounding them. The shedrow is bisected in the middle with a wider

---

1. See Appendix 2

aisle. All traffic moves counter clockwise around the shedrow and walking horses is referred to as '**turning left**'.

In the open area of the H are located the hot walking machines (if there are any), the dumpsters for manure and an area where the employees park their cars.

Duck walks me over to the racing office to get my groom's license. This is the nerve centre of the track, where races are entered and all the officials have their offices. The Ontario Racing Commission is the branch of the government that protects the public from unscrupulous practices in racing. They license the backstretch personnel as well as ensuring all the racehorses are sound and healthy on the day of their race.

I fill out a form, am photographed and finger printed feeling much like a criminal. While the official rolls my inky fingers, he comments on how soft my hands are for a groom. I know full well that with the mucking, washing, poulticing and doing up horses with Absorbine, my hands will soon toughen up. I am handed some paper towels to wipe the ink from my hand and my photo ID is made up and handed to me. Officially, I am supposed to wear the badge whenever I am on the backstretch. Mostly, I keep it in the car.

Mornings begin at 6 am. I give my name at the gate and drive to my barn. Each groom has four horses assigned to their care which is a full work load. The first order is to pull the feed tubs and bandages, checking the legs for heat and swelling. I start mucking the first stall, tying the horse to the wall. The **exercise rider** drops the tack off for the first horse to be galloped. Each horse gets a perfunctory brushing (called a **knock off**) before tacking. The girls teach me the race track method of **picking** all four **feet** out from the same side.

Once the horse is tacked up, I lead him out into the aisle. The rider is **legged up** and led down the aisle as the rider tightens his girth and knots his reins. Then the horse is led to the end of the barn and turned loose. While the horse is galloping, the stall is mucked. Clean straw is carried from the end of the barn to bed the stall, the water pail is dumped, scrubbed and refilled. When a horse returns from the track, he is bathed and a **cooler**, which is a large wool blanket, is thrown over the horse. Then he is tied to the barn's hotwalking machine.

This routine, with minor variations, is repeated for each of the four horses.

There are still legs to be scrubbed of the **poultice** which was applied the day before. The trick is to find a time when our handyman Jack is less busy (he thinks he is always busy) so he can hold the horse while the clay is washed off. Some of the horses are **ponyed,** particularly the older, sore horses. To pony a horse is to have him led by a person riding another horse. All the **lead shanks** at the track are leather with a chain attached. The **chain** is wrapped around the nose and **halter** which the pony boy holds in his right hand while guiding his mount with his left.

By 9 am, most horses are finished their morning gallop unless they are to be **worked** on the grass. About this time the coffee truck appears. Almost everyone lines up to buy the swill heavily laced with chicory that masquerades as coffee and overpriced food served up by a surly Greek. I quickly decide to pack my own Thermos of coffee and some sandwiches. There is no employee lunch room so I squat by my stalls to eat, avoiding horses still being hand walked. Now I begin to 'do up' the horses.

Before the racehorses gallop, they have a quick brushing to make them presentable. Now they receive a more thorough going over. Because of their fitness, **thoroughbreds** are thin skinned and do not tolerate stiff brushes and rough currying. A rubber glove may be used as a curry followed by a soft brush to polish the coat. The mane is laid over neatly with a wet brush. Tails are hand picked and combed with fingers. If a horse has worked or raced, the feet are packed with clay to draw heat and soreness, then covered with a square of paper to keep the clay in. Hooves are oiled around the coronary band to keep them moist. A rub rag is run over the body to pick up any stray dust.

This late in the year, almost all horses' legs are bandaged. Some legs are vigorously massaged with Absorbine Liniment. Others are poulticed with Uptite, a white clay that draws heat and firms up puffy fetlocks, which we call ankles colloquially. The girls teach me how to take a handful of clay, dip it in warm water and rub upwards against the hair to adhere the clay to the leg. After making a clay cast, the material is covered with plastic (to keep it moist) or paper (to tighten it). Quilted cotton or doubled thickness of fleece is wrapped around the legs and snuggly held in place by cotton bandages. These bandages are white (called lily whites) or red and bought in great rolls. They are then pinned vertically with bandage pins. A mixture of cayenne pepper and water is smeared over the pins to keep busy teeth from ripping bandages off over night. I often forget the pepper, holding the pins in my mouth when doing the bandages and burn my lips.

As each horse is done up, his halter is removed and hung neatly on a hook centred on his door where his brass nameplate faces out. By 11 am, the horses are done up and feeding commences. A huge tub of feed is hung in the corner of each stall and the **webbing** secured across the front of the stall. Haynets are hung up in the open doorway; brushes and saddle towels are hand washed and set out to dry. Halters are polished with saddle soap and Brasso. The aisle is raked. Grain is set for the afternoon. Now, unless there is racing, we are free to go until afternoon feed. Our expensive animals are now immaculate. The grooms? Not so much. I am filthy.

Afternoon feeding starts at 3:30. This is the main feeding. While the horses eat their grain, the groom picks out the stall of any manure and wet spots. Water pails are filled for the night. Saddle towels are folded and put in the **foot locker**, along with the dried brushes. The aisle is raked again.

As swing groom, my job includes covering for Jack. Jack's job seems easier, consisting of putting up to four horses on the hot walking machine to cool out and to set the timer for each horse. Horses that galloped need to be offered water every few minutes from a bucket. I hold the horses for their baths, trying to avoid being bitten by the colts. (True to character, Jack's solution is to take a swing at the offending horse, usually while the trusting groom is squatting down to wash feet). One of our geldings on the hotwalking machine habitually stops whenever he feels like it. The machine squeals ominously while the other horses obediently wait for pressure on their halters to start again. Tossing rocks at the gelding convinces him to resume walking.

One day while doing this job, all hell breaks loose. One of the horses rears up, getting his foreleg over the rope connecting him to the hotwalker. The horse throws himself madly, tearing the skin on his leg.

"Where's the knife?" Linda screams at me.

Knife? What knife?

There is supposed to be a knife nearby to cut the horse loose in just such an emergency. Of course, Jack has borrowed it for some mundane task and has forgotten to return it. Eventually the animal is freed but not before there is some serious damage done to his leg due to a rope burn. This should not have happened and everyone in the barn is upset. The anger is focused on me as much as Jack.

Dean decides to try a new product, Animalintex poultice, on the leg. Within days, the leg is miraculously better. I am still not forgiven. For what, I am not sure. I feel very much like a useless appendage at times. I try to do my job to the best of my ability but I always seem to fall short. When I arrive in the mornings, the others are already at work and I feel like an ill fitting jigsaw puzzle piece. While Tommy seems fine with my work, I feel I do not measure up for Dean.

Beasley Farms breeds and raises their own race horses. For a couple of weeks, the Beasley girls, as we are called, assist with the breaking of these **yearlings**. Dean drives us north of the track to a pretty farm tucked away in the Caledon hills where these horses are raised. We teach the 'babies' as the yearlings are known, to carry a saddle and bridle and to be ground driven. **Ground driving** consists of hooking a pair of long **lunge lines** through the stirrups to be fastened to the bit. The driver (us) takes the other end of the lines and sends the horse forward at a walk, teaching the horse to turn left and right and to stop. Duck and Linda have done this in previous years while Sam finds this more difficult. Since I learned this skill this past summer, I have a relatively easy time of this. What I observe from this is the importance of good breeding. One of the fillies is **inbred** and she 'washes out' with this simple exercise. **Washing out** means sweating profusely due to nerves. None of the rest of the babies are the least bothered. After a week of ground driving, the babies are **backed** (ridden for the first time). Several days later, they are shipped into the track.

This late in the season, any horse that is unsound for any reason is shipped home. There is not enough time for the horse to recover and be fit enough for racing before the end of the season. The number of horses at the track gradually decreases from August onwards, leaving stalls available for the stables like Beasley who want to bring their babies in and expose them to the track environment.

I am given a few extra days of work rubbing yearlings at the track. The 'babies' always go out in pairs to settle them in the unfamiliar setting. Jimmy, our jockey, is still riding half of the yearlings. The morning the babies are due to go out for their first gallop, an exercise rider shows up who I have never seen. What I do not know is that he is strung out on drugs. Before I touch the rider's leg, he leaps clean over the animal and lands in the straw beside her. I am not sure who is more surprised, the filly or me. That was pretty damn athletic but the filly could easily have kicked the rider. Fortunately, he disappears after a few days and an older, more reliable freelance rider takes his place.

This man seems older than many of the exercise riders. He announces himself on the shedrow with a cheerful whistle, is quick and business-like and greets me in a friendly manner.

A couple of days later, he invites me to a dinner party at his home, along with Linda. I rarely get party invitations and I decide to accept. How different the racetrack is proving to be than my cliquish high school was.

The apartment is around the corner from where I am living and I enjoy my evening. I am made to feel very welcome and my host invites me to drop by any time.

After a couple of weeks in training, the babies go home. Several of the older horses have already gone to the farm for the winter. The barn quickly downsizes. Sam is demoted to swing groom again and I am laid off for the winter. It is now late November with training to resume in February.

Having no income, I move back to my parents' home. I sit idly at home, watching late night television as the weeks pass. Occasionally, I drive back to Toronto where I visit my new friend the exercise rider at his apartment. I am warmly welcome there.

By mid February, when I have not heard from the trainers at Beasley Farm, I decide to drive up to the track and check things out. I find the horses have shipped in and the girls are hard at work. There is a new girl working too. Neither trainer is on the shedrow. My questions elicit vague replies. No one will look me in the eye. Finally Dean appears and tells me I have been replaced, that there is no job for me here.

I am secretly relieved that I will no longer have Dean's critical eyes looking over my shoulder. I quickly formulate a new plan. The track is just beginning to fill up with a demand for skilled workers.

Being already trained, I can go elsewhere. I head for Jimmy's trailer.

Chapter Three

GREENWOOD SPRING MEET

Racetrackers are a different ilk of people. While they would die cooped up in an office or factory, they relish being outside in any weather. They are often unkempt in appearance but there is an authenticity to them I find refreshing after so many years of the pretensions of school.

Jimmy's spartan trailer office is just as bare as I remember it. Although still early in February, the back stretch is already a flurry of activity. Vans are pulling up to the gate shipping racehorses back in with stable help flocking back to their barns. Horses are already being ridden around in the shedrows beginning their conditioning.

Jimmy has more flexibility of where to place me now that I am experienced and the track is hiring again. He tries to give the better positions to the girls, where they are not as likely to be taken advantage of (I will learn later of the rampant sexual predation that takes place from which I have been protected). After a few phone calls, Jimmy drives me down to Barn 3A and introduces me to Bobby Fischer.

Bobby is an ex-jockey who as a youth specialized in riding at the half mile tracks in Ontario. Alcoholism eventually ruled his life and the man was lucky to walk away from a couple of serious car accidents. Now a recovering alcoholic, he trains horses for a living.

In the 1940s and 50s, racing in Ontario consisted of tiny raceways often tucked into odd corners of a city. Each meet consisted of two weeks racing over half mile tracks, often at local fair grounds, known as a 'leaky roof circuit'. Horses and grooms alike shipped from one poorly maintained track to another by train. There were no proper living quarters for the grooms; they simply unrolled a sheet of linoleum in an unused stall and they threw their sleeping bags down.

Racing enthusiast and multimillionaire E.P. Taylor set out to change racing in Canada to bring it in line with the best American raceways. He gained control of the Ontario Jockey Club, which systematically bought up and closed down these little raceways. To implement Taylor's grand scheme, The Jockey Club bought 680 acres of farm land on the outskirts of Toronto to build a grand new facility. The result is Woodbine, a large and modern racetrack which aspires to greatness.

It takes me a while to understand the general workings of racing. The racing secretary is the official of the Ontario Jockey Club (who run the Ontario thoroughbred racing circuit) that 'writes' (compiles) the **condition book**. For each day of racing, the book lists the races that are offered. Each race will state the distance run; the prize money, the sex preferred; number of races previously won. There is one more race listed on the card (daily racing quota) than will actually be run. The race with the lowest number of entries (unless it is the feature race) is thrown out.

Entries are made by 1 pm two days before the race day. The racing secretary makes **the draw**, which determines which horses are included. A second draw determines their post positions. When a race overfills, up to four other horses are listed as **also eligible**. If a horse is scratched in the next 24 hours, horses from the also eligible list are moved onto the race card. The draws are witnessed by trainers, owners and agents, keeping the whole process transparent.

Still, the backbone of racing is its **claiming** races. The barn I am heading to is one of these stables where the horses are cheap and run often.

Bobby's barn is a considerably different from Beasley Stables. As a public trainer, Bobby has a variety of clients. There is a high turnover in staff and horses. We run them often, patching them up to stay sound. Bobby supplements his income by owning one of his own racehorses. His full complement of staff includes four grooms, four **hotwalkers** (Bobby does not own a hot walking machine), two exercise riders, the pony girl and Bobby's wife Shirley. The barn is always a chaotic hub of activity. I am hired on the spot, given a pitchfork and set to work mucking a stall.

"Hi. Are you the new groom? My name is David but everyone calls me Fourfoot." The young man says this all in one breath as he waves to me from the back of a horse he is shedrowing. Fourfoot looks like he cannot be older than twelve. I learn later he is twenty and hopes to become a jockey.

Even younger is Charlie, a pimply faced but pleasant lad who is walking hots. Bobby is giving him a chance to start exercising horses. Paula, the returning groom and Rita, the pony girl are in their mid twenties. One of the other grooms is older, the other younger than I. All of our hotwalkers are hired off the street with no experience (I am sure some of them have never even patted a horse before). They are mostly high school dropouts and this is their first job. Rounding out the staff is Bobby's wife Shirley, who sets feeds, rakes aisles and generally acts as den mother to the girls in the barn. Unlike Beasley Farms, there are no days off in this barn, resulting in a revolving door of help. Yet I find the people working here very friendly, in contrast to my previous work experience. Here, I finally feel as if I belong.

This early in the season (winter actually), the horses are not yet assigned to grooms. Our mornings start later at 7 am due to the lack of light outside. Each day I am told which horse to get ready and which stall to muck. Since the stalls do not have lights in them (Beasley Farms must have installed them at their own expense), I hang a trouble light (the kind that a mechanic uses under the hood of a car) on the screen as I work. The aisles are quite dark too. The outer walls of the shedrow are just plywood half walls. In the winter, canvas or plywood covers the upper half, making the shedrows dingy.

Only a few years before, women were a rarity on the backstretch. By the late 60's, however, the effects of Women's Liberation are felt even here. As women begin to break out of their traditional roles, the racetrack is one male dominated industry in which women begin to infiltrate in numbers. Although about one quarter of the backstretch is now female, there are still only two washrooms in the entire backstretch assigned for us. Both are too far for us to walk to during training hours. Shirley gathers all the girls in the barn (including other stables) in mid morning to use the men's toilets. We take turns guarding the entrances to use the facilities. Some of the men are so uncouth that they will not respect the girls but barge in anyway. Many men resent the girls working on the backstretch, claiming they have 'ruined' the track and are stealing their jobs. Since the girls are willing to work harder and put in longer hours than many of the men, the girls are popular with trainers. There is also resentment because some girls sleep with their trainers. These girls use sex instead of capability as a way to obtain status for themselves.

Like many of the young girls on the track, I am also dating an older man. He is the exercise rider who invited me to his dinner party last fall. When in Toronto over the winter (I kept my apartment), I visited him a few times and we started dating. the Boyfriend drops by the barn from time to time but we do not let on we are seeing one another. The small town nature of Woodbine means everyone is familiar with everyone one else and I dislike my private life to be on parade.

The girls at Beasley have taught me well. I am now competent with putting on exercise bandages, doing up a horse in Uptite poultice and have even mastered the tricky **spider bandage** used to do up knees. By the end of February, Bobby assigns the grooms their horses. Training hours are moved back to 6 am. The season has started in earnest. I am pleased with my roster of horses.

At the far end is Tam O'Shanter, a nondescript bay gelding of dubious ability. His only talent is getting into trouble. When I lose an expensive leather glove in his stall, I tear the stall apart but cannot find the glove in the thick bedding. Next morning, as Shirley is setting feed, she asks about what she found in the bottom of the feed tub. There, chewed to a pulp and spit out, is my frozen remnant of a glove. It is now garbage.

Tam O'Shanter, usually called Tammy, also develops a trick of turning himself loose from the hotwalkers. When the kids are watering him off, the gelding takes the opportunity to rub his nose on the pail, unhooking the snap. Next time the hapless kids pull on the shank, the chain slides harmlessly off his nose. Then Tammy gallops away and has to be chased about the barn. After turning himself loose in this manner several times, we twig to his trick and warn the hotwalkers not to let him rub his face. Tammy also drops his nose to the floor when I am bridling him so I have to bridle him literally on my hands and knees. He tears his bandages off regularly. He is like that kid in the classroom when you were in public school who made faces when the teacher's back is turned. The one with the pea shooter who cannot pass a test if his life depended on it. Last year, Tammy **broke his maiden** as a two year old. With the lack of effort he puts into racing, he will probably never win again.

In the next stall is a small, dark two year old filly named Magpie's Ghost. She is the sole horse of her owner, a genial man called Jack who soon becomes our favourite owner in the barn. Every weekend, Jack comes to see the filly and he revels in the life of the track. Jack loves hobnobbing with all the help in the barn, unlike so many snobby owners.

Magpie, or Maggie as we call her, proves to be a real character. She is feisty and intelligent. She frequently pins her ears back to her head and bares her teeth, intimidating everyone around her. With her fuzzy topknot, she is a brat who likes to throw a hissy fit to get her way and is pretty free with her heels. When she bounces down the shedrow with a rider on her, the shedrow parts like the Dead Sea. Except when she comes into heat. Then she becomes sweet and cuddly, unless you kiss her nose, causing her to squeal fiercely.

Of course, almost everyone then plants a kiss on her. When I lead her out for Bobby to ride, she tiptoes down the aisle, peeing all the way. Bobby says he has only seen one filly worse in heat than her. In her stall, she and I soon come to an agreement. I train her to stay on one wall until I step back and cluck, when she obediently moves over to the other side of her stall. I am safe around her and she becomes one of my favourite horses of all time.

The third animal in my care is Bobby's own horse, a 6 year old gelding named Pass to Win. His barn name is Winnie which morphs into Pooh Bear, as in Winnie the Pooh. He is strongly built but is a crippled campaigner. Pooh Bear is almost always **good for a cheque** (finishing in the top four) and is Bobby's consistent bread winner.

In this horse comes one of my great lessons in horsemanship. Pooh Bear hates me. Nearly five years before, at my very first job with horses, this same chestnut colt was one that I occasionally handled. When leading him to his paddock, the horse would rear and refuse to lead. When he reared up, I smacked him with the lead shank to make him mind. Then I backed him up all the way to his paddock. It was what everybody did. I did not understand that the reason he acted up was because his tender feet hurt on the gravel road. A horse's memory is second only to that of an elephant. Five years later, he still holds a grudge. He pins his ears to greet me and kicks when I brush him.

On the other hand, the gelding truly loves Bobby. His ears swivel forward at the approach of his trainer. I always know when Bobby has arrived at the barn by the smell of peppermints on the horse's breath. Fourfoot tells me that once, last year, Pooh Bear got loose. He simply raced around the corner and ran straight up to Bobby.

Pooh Bear is also a **stall walker**. It is a nervous habit in which the horse paces around his stall ceaselessly. When I arrive in the morning, there is already a track of mulched straw and manure packed in his stall. He is even worse on race days. With stall walking, the horse is said to '**leave their race in the stall**'. It is also incredibly hard on their legs, wearing joints out prematurely.

Over the course of the season, through trial and error, I work out a routine that controls the stall walking. I discover if I arrive just a few minutes before 6 am, the stall walking has not begun. Taking a full length of binder twine, I tie his halter to the screw eye the water bucket hangs on. Simply stopping him from walking but still allowing him to look out and move about the stall arrests his behaviour. I leave him tied during the day (he never tangles himself, fortunately) and untie him at feed time. The strategy works perfectly and his race performance improves. The horse's state of mind is critical to his racing efforts. A happy horse runs better.

My fourth responsibility is a delightful **dark bay** stallion named Captain's Resort. He is robust, about 15 **hands** high and beautifully put together. For his first two days back at the track, he acts studdish, prancing proudly and screaming at other horses, having bred two mares the year before. Then as if he remembers that he is at the track, he becomes a perfect gentleman. Unlike most stallions, I do not need to be on constant guard when working around him. Linda, from Beasley Stables, says when she rubbed him the last year, his barn name was Teddy. My nickname for him is Capitan.

Credit: *Michael Burns*

In this horse comes one of my great lessons in horsemanship. Winnie hates me. Brian Swatuk up.

Capitan has a unique habit. Each morning as I am saddling him for Bobby, he stretches out his front legs and bows like a dog to stretch his back. He does this four times each morning, without fail. It always brings a smile to my face and Bobby's as well.

We train most mornings, unless the track is frozen hard. In that case, the exercise riders 'shedrow' the horses, riding them around under the overhang of the barn. The riders trot wherever they can find any space to do so. If a horse is **worked** the day before, they are hand walked for half an hour for some light exercise. Horses are also hand walked for half an hour after galloping. In mid morning, the shedrow is completely full of horses turning left.

When we say a horse is galloped, in fact it is only ridden at a sedate canter. As the horse gets a **'bottom'** (certain level of fitness), the horse begins to **work**, usually once per week. Older horses may be galloped two miles per day. Two year olds take 2-4 months to get fit, barring no set backs but older horses can be ready to **work** in a month. The first works are short, usually one quarter of a mile and the distance increases as the horse gets fitter.

While the days are much alike, little differences set them apart. Some days the horses return from the track with frosted whiskers from their breath freezing on the face hair in the cold morning air. One morning, a snowfall produces such lazy big snowflakes that the horses come in looking like instant **Appaloosas**. Despite our hectic mornings, we take the time to notice each feature of every day.

It rains constantly this spring. The track becomes increasingly muddy and deep. It is too cold to bath yet. Bobby teaches me how to tie up a **mud tail**. The tail hairs are parted in half, wound around the tail bone with the last hairs braided together and tucked securely in. This keeps the tail mostly free of mud and sand on wet days of training.

Maggie, however, will have no part of having her tail done up. She kicks so violently that by the time she is half way around the shed, her tail is down. Bobby tells me to leave it that way. It takes a long time to brush it clean every morning when she gallops in the mud. Since I cannot afford rain boots, my jeans are full of mud up to my knees by the time I leave each day. I hand rinse them, hanging them up to dry in the bathroom at home.

By the third week of March, racing begins. Bobby lets the grooms know if he has entered one of their charges. By feed times, the '**overnights**' are out and we are responsible for picking up a copy in a rack outside the office. The overnight is a page listing all the horses racing with their post positions, including the also eligibles.

The night before a race, the haynet is removed from the stall. This practice, called **drawing**, prevents the horse from filling up and not turning in his best race. The next morning, the horse is hand walked for half an hour, usually first thing while the hotwalkers are not busy yet with horses returning from their galloping. The stall is mucked but not bedded to discourage the horse from eating the clean straw.

By mid morning, the Racing Commission vet shows up. The horse is jogged (trotted in hand) for soundness and his **lip tattoo** is checked against his registration papers. The Jockey Club makes every effort to ensure the horse is sound and that there are no **ringers** used (lookalike horses). Any horse **riding in the transfer van** is fed their lunch grain early. Water, however, is withheld for several hours before the race.

There are three racetracks in our circuit. The racing season begins at Greenwood track, located in the opposite corner of Toronto. Since we are stabled at Woodbine, we begin boarding the transfer vans at 10 am. The loading ramp is situated in a parking lot beside the track kitchen (restaurant). The **stall man** calls us down, one van load at a time, starting with the earliest races. I present a somewhat comic figure, trudging down the road, race horse in one hand and metal lunch box in the other. We lead our horses up a wooden ramp and back them into a tight stall. We snug up their tie chains on either side of the halter to prevent fighting. There is usually a total of twelve horses in the van. The grooms and hotwalkers ride with their horses, keeping the animals out of trouble. It is a restless ride with our hungry horses. Usually, we keep their shanks on. Bobby takes my brushes, water pail, cooler, bandages and hay net with him. He and Shirley meet me at Greenwood.

One particular race day is etched starkly in my memory. It is a dreary March day. Pooh Bear is in the last race. I sit on an upturned bucket outside the stall in the **receiving barn**, wrapped in a cooler as the day drags on. Race horses are never left unattended on race days. It takes only seconds for anyone to slip into a stall and drug the animal to prevent your horse from winning. The cold seeps into me and and I am miserable. I ration my coffee from my Thermos to make it last as long as possible. Shirley gives me about a half hour reprieve to warm up in the kitchen, where I buy another coffee and watch a race run on the closed circuit TV.

Since I am in the last race, I am almost as bored as the **hotwalkers** turning left. Hot walking is so mind numbing that the young kids often forget to watch that the **cooler** stays in place as the horse cools out. A favorite trick is to watch the blanket fall onto the shedrow and scoop it up, then wait until the hotwalker is far down the shedrow to yell, "Watch your cooler, kid."

The poor kid turns around to find his cooler has inexplicably disappeared. He is, of course, looking close by him. Everyone else knows exactly where the blanket is. This provides an entertaining diversion in a long day of waiting.

By the time I head over to the race with Pooh Bear, white fluffy flakes are beginning to float down. The race is uneventful and Pooh Bear finishes third. An hour later, by the time we have cooled out and are ready to head home, a winter storm is well under way. Everything is now soaking with the wet snow. The wooden ramp is slippery. The van driver helps by steadying the horses by their tail as they load. As these are the older campaigners, we should be quickly loaded and on our way.

Unfortunately, an huge older horse bangs his hip as they attempt to manoeuvre him into the van's stall. He backs down the ramp and then refuses to load again. For half an hour, the groom and driver attempt to reload the horse before he is finally coaxed back on the van. Darkness falls. The van is closed up tight, hay nets are hung and we start for Woodbine.

Now, however, we are in the midst of rush hour traffic. What is normally just a half hour trip becomes a stop-and-go nightmare as we inch along through downtown Toronto. My clothes are soaked through. Even with the heat given off by the horses in the enclosed space, I have never been so cold. The feeble light in the van allows us to see the horses and each other but no one feels like being social.

The hungry horses tear happily into their hay. They have not had any hay in over twenty four hours. This food makes them much quieter than on the way down. With the floor wet, there is nowhere to sit and the grooms must either stand or squat. When we squat, the horses pull great hanks of hay out of their nets, snorting and showering us with the fodder.

We unload the horses in the dark. We've arrived much later than expected. I should have been home hours ago. I fling the hay net over my shoulder, grab my lunch box and lead my horse back to the barn. There is an occasional light on in a barn, otherwise, Woodbine is deserted. The stall, which should have been freshly bedded, has been forgotten. I feed Pooh Bear his **bran mash**, hang the hay net and proceed to muck and bed the stall by trouble light. I **pack his feet** with clay, poultice his legs and wrap them.

Finally I drive home to change into dry clothes. It has been a long day. Yet before this, I was just marking time, waiting for my life to begin. Now I do not have to justify being around the horses. People here do not care what your past is. All that matters is today. I blend into the rhythm of the track.

Chapter Four

THE LITTLE BIRD OF A FILLY

Spring is a time of great optimism in racing circles. All of the horses are earnestly in training. The two year olds are still an unknown quantity, giving great hope to their respective barns. With time off to rest, the old campaigners are still sound. The warmer weather has coaxed the green grass up. The canvas that encloses the shedrows has been taken down. This leaves our barns are open to the sunshine, adding to our enjoyment of the weather.

My filly Magpie is now fit enough to race. The first two year old races has been written. One afternoon, I lead my two year old over to the test barn to be tattooed. Her papers are checked to see that her description matches her actual appearance. The tattooing device is made of blocks of steel needles which form a letter and numbers. The letter, which is E this year, corresponds to the year of Maggie's birth. A burly man applies a rectangular clamp to her top lip and curls it back to expose the inner lip. Each needle block is firmly pressed into the lip. Black ink is smeared across the inner lip and the result is photographed. Maggie has been stamped with her permanent racing number, corresponding to her registration papers.

Maggie has also been **okayed from the gate** (she knows how to break from the gate) and made the time sheet (turned in a suitable workout). She is officially ready to race. Bobby finds a race and enters her. Two days later, I set out for Greenwood.

When I am called to the transfer van, Bobby comes down to help load her. He takes the shank from me and tries to coax her up the ramp. She does not budge. Next, two of the van drivers try to lock hands behind her to push her on. Maggie pins her ears flat against her head and threatens to kick anyone except me who comes close to her hind end. Exasperated, the burly men confer for a minute or two. Then they turn the filly around and carry her on backwards. Her front feet are several inches off the ground. I swear she is smiling.

She is so excited by the trip that she develops hiccups. In the **receiving barn**, she bangs her stall door with her front leg, paws up the straw and continues to hiccup. The hiccups continue all afternoon long, much to the entire barn's amusement.

The track is **greasy** from a recent rain. As soon as Maggie leaves the starting gate, it is evident she cannot handle this going. She is literally swimming in the mud. The rest of the field stretches further and further out in front of her. She is beat by an embarrassing 31 lengths and is officially listed as '**outdistanced**'. Oblivious to her humiliating finish, Maggie cools out fine but is tired. Bobby has already left when I load her for the trip home. She obediently follows me onto the horse van with no hesitation.

Two weeks later, Maggie is entered for her second attempt to race.

That day, I am not even finished doing up my horses when called to the van. Fourfoot is assigned to finish my work. I show Fourfoot how to bandage Tammy. The horse paws continually while I am trying to wrap him. With practice, I have learned to pass the bandage around the leg while the foot hits the ground. It takes no more time than usual to do him up. I remind Fourfoot to put some red pepper on the bandages to prevent Tammy from chewing them off.

Maggie's second race is almost a repeat of the first, minus the hiccups. Again Bobby attempts to load her and again the burly truck drivers back her onto the trailer. Her losing margin is not as dramatic this time. Once again, I load her for the trip home with no issues.

Next morning, I find my entire jug of red pepper and water completely empty. Tammy's bandages, however, are so encrusted they could stand up by themselves.

"Fourfoot! How much pepper did you use?" I exclaim.

"Well, he didn't bloody well touch them, did he?" Inwardly, I sigh. I am never sure what will happen in this barn when I am not there to monitor my horses.

Maggie has one more race at Greenwood. This time Bobby leaves me to load Maggie by myself. She obediently follows me up the ramp. She is less anxious before her race and the track is fast. She puts in a decent performance to finish fourth. The filly is sent to the test barn to cool out and have a urine sample taken.

The test barn is a small barn centrally located at each track. It consists of four stalls, a wash aisle, a lab and a shedrow around the works. The winner and at least one other horse from the race (steward's decision) are sent to the test barn. Since it is heated in winter, every horse is bathed from the barrels of warm water set for us. Each horse is assigned one of the different coloured pails of drinking water that are set around the perimeter (this prevents cross contamination with other horses. Fresh water is provided for each animal). When the horse is finished drinking (they usually will drink 5 or more gallons of water), the lab person assigns us to a stall and enters with a plastic cup set in a long handle. They whistle until the horse urinates and catch some of the urine in the cup. The cup is taken into the lab, affixed with a label and sealed in the presence of the groom or hotwalker. For security purposes, the label is **cut irregularly** and the urine sample is sent out to a lab to be tested for banned drugs.

Maggie appears to be limping as she cools out and she is still thirsty. After I return to the receiving barn, she drinks another full pail of water. On the van ride home, she keeps trying to balance on her toes and take a pee. Finally, we leave the highway and stop at a red light. Maggie has an enormous pee. When every one in the trailer gets up and stares at her, she tries to hide her head in shame. I find it funny that a horse can be embarrassed in this way.

Next morning it is confirmed that Maggie has bucked shins, the scourge of young race horses. This condition is almost exclusive to race horses. As horses begin to get fit, the cannon bone below the knee starts to thicken in response to the stress of galloping. It first lays down a thin matrix of bone, which grows stronger in time. Under the stress of running fast, this new bone growth can fracture. Bucked shins is very painful, yet Maggie had continued to run despite the pain. That showed real heart and class.

Maggie is sent to a farm for six weeks to heal. When she is due to come back in June, Bobby actually sends me all the way to the farm to load her onto the van. Despite the warm weather of June, the filly is still fuzzy with her thick winter coat. She also gained weight on her lay up. She is so chubby she looks as if she has never been fit. Bobby brings me a fine toothed hacksaw blade to drag across her coat to shed it out. As the long fur is pulled out, there is almost no summer coat beneath it. I can see the bare skin underneath. Eventually, her sleek summer coat does grow in.

Conditioning starts again. By early July, she is deemed fit enough to race. Bobby tries another jockey, Billy on her. Bobby tells the jockey she might '**come up short**', that he (Bobby) is giving her a race. Maggie does runs well for the first part, then gets tired and fades to last. Bobby greets the jockey after the race to ask him about Maggie's performance. Billy, the jockey, says she just came up a bit short, that he folded up and let her ease back. Then he asks Bobby to give him the ride again.

Back at the barn, Jack, her owner, tells us how he squirmed when he overheard someone in the paddock point to Maggie and say, "Look at the fat little thing with the short legs." As her coat sheds out, the baby fat is soon replaced by hard muscles.

A couple of weeks later, I am giving Maggie her final polish before taking her over for her next race. John Doe, the hotwalker, is watching and comments, "Better get her ass shiny 'cause that's all they're going to see." Our whole barn is keyed up about the race.

It is such a gorgeous summer day. In the paddock, I lead Maggie into stall 2, turn her around so she faces out and take up my position in front of her, a rein in each hand to control her. Jack is there with Bobby beside me.

Credit: *Michael Burns*

Better get her ass shiny 'cause that's all they're all they are going to see

The valets spew out of the jocks room with the saddle cloths, postage sized jockeys' saddles and **weights** draped over their arms like a maitre 'ds at a fancy restaurant. The valet saddles from the right side of the horse, places each item of tack on the animal while the trainer secures the tack from his side. Special elastic girths are used for racing which can easily pinch the horse. The valet attaches the girth on his side, stretching it down before handing it to the trainer to be done up. Then the overgirth is laid on top of the saddle and both sides pulled down simultaneously. It is fastened underneath the horse, between the forelegs.

The last step is for the trainer to stretch each of the forelegs out to prevent the skin from pinching (and causing the horse to buck). By now Maggie is an old hand at this. As Bobby reaches for her first leg, Maggie obliges by smashing the leg forward. He points at her other leg and she does the same. Bobby looks at me and smiles. I have never known another horse to do this.

The jockeys now swagger out of the jocks room, secure in their status as leading actors in this play. The first time I **paddock** a horse, I nearly laugh aloud at them. Dressed in the colourful silk shirts, helmeted heads and tiny hips clad in white breeches making their hips look smaller, the jocks look doll like, smaller than their actual size. Billy joins us for last minute instructions.

"Just take her back a bit, off the pace, then move her up at the quarter pole," Bobby advises. "She should be right there."

With the call "Riders up," Billy offers his bent knee to Bobby and is boosted into the saddle. I switch my grip to the chinstrap, leaving the reins free for the jockey to knot them. I join the parade behind Alfie, the head **outrider**. The bugler finishes his call and rides back past us into the paddock as we walk onto the track. I hand Maggie over to the pony boy and slip into the narrow space by the finish wire where the grooms stand.

Maggie's race is text book. She breaks well, settles back and swings into contention at the top of the stretch. She duels down to the wire and wins by half a length. I am screaming my head off the whole length of the stretch. As the jockeys pull up on the backstretch and turn around, I walk out onto the track to await my filly. Bobby quickly joins me, giving me a congratulatory hug. Jack and his family appear from the grandstand for the win picture.

"Number two, test barn." An official behind me barks.

Billy rides the filly up to me. The track photographer has emerged from the grandstand and sets us up for the victory photo. The race is confirmed official and the payoffs are posted on the tote board. Maggie has paid off at a whopping $114.00 for a two dollar bet, the highest payoff of that year. On paper, she looks like a poor choice, being last three times and fourth once. However, the first two races were in the mud, which she could not handle, then my gutsy filly still raced well on her bucked shins despite the excruciating pain. After her layoff, she came up short and Billy let her finish last. Her win is no fluke.

When Maggie returns from the test barn, our barn explodes with joy. Everyone fusses over the filly. Jack throws an impromptu party. It turns out that no one in the barn bet on the filly. Even Jack only had a modest sentimental bet.

Actually, somebody did bet her. Billy had a hundred/hundred on her (to win and to place). I heard he got married later that year. It must have been a tidy nest egg on which to start a marriage on.

Credit: *Michael Burns*

Maggie has paid off at a whopping $114.00 for a two dollar bet

Chapter Five

Number 47

I've become part of the crazy quilt that makes up the backstretch. This place is a hodge podge of colourful characters. The well educated are indistinguishable from school dropouts, wealthy businessmen defer to the trainers and jockeys and there are plenty of guys around with petty criminal records.

The community of the track is incredibly accepting of everyone and I find it so easy to make friends among the grooms. I fit with people here instead of feeling like the social pariah I have always been. For me, the warm summer days roll easily into one another like wet paint that smears on canvas.

At Bobby's barn, the hotwalkers come and go. With no days off, these kids quickly wear out. At one point, our most experienced hotwalker has been with us just one month. Grooms, too, are very expendable, one day here, next week gone. This parade of help means we are always training new help. My time at Bobby's is teaching me how important staff are to racing success. In the better barns, experienced help stays for years and the consistency of the staff results in a better win average.

I get frustrated by having to constantly train new people. When I prepare for the return of my horse from the track with a cooler and shank carefully laid out opposite the stall, I find them snatched away right from under my nose by a hurried hotwalker. Bath water, too, vapourizes if I do not guard it fiercely. Saddle towels walk down the aisle to be found in other grooms' boxes. Bandages end up on horses that I am not responsible for. Only by snarling like a bear if anyone so much as looks at my stuff does it stay put.

When the Greenwood meet ends, all of the horses ship home to Woodbine. Now the main track opens for galloping in the morning. There is now a choice of four locations to train – training track, main track, sand ring and back field. Since we are closest to the training track, we continue to use it most of the time.

We are bathing every horse in the morning now as they come in from the track, making the grooming after much easier. However, most of the horses are done up in bandages, either with alcohol or Uptite. It is easy to spot the grooms on the track. We are the ones with untidy hair pulled back in a pony tail, a hoof pick in our pocket and Uptite smeared on our dirty jeans.

I am seeing more of the Boyfriend these days. He takes me clothes shopping, buying me three pairs of work pants that would cost me the best part of a week's wages. I am touched by his generosity. Being in love causes me to glow and some of my friends comment on this. Still, within the closed world of the backstretch, I keep my private life a secret.

I am learning to keep my mouth firmly shut. Talking too much about your barn lets others know what is going on in your barn, which horses are ready to win, which horses are breaking down. This can result in anything from not being able to cash a handsome bet to the more serious consequence of having a valuable horse claimed off you. Empty stalls mean laid off workers.

Bobby, my trainer, has one peculiarity that surfaces now. On race days, whenever we run a horse, he demands that all the grooms stay all day. In most barns, just the groom and one hotwalker are all that are required. When we race at Woodbine, there are four grooms, the hotwalker, Bobby and his wife Shirley, Rita the pony girl and both exercise riders just to run one horse. I chafe at the restrictions on my time. Since claiming horses are Bobby's bread and butter and the two year olds are just coming into form, we now race several times a week.

Bobby also nixes trading off afternoon feeding with another groom, even one in our barn. Cassie and I have been in the habit of doing this. By coming in fifteen minutes earlier and staying fifteen minutes later, the afternoon feed is covered and the other person has from 12 o'clock off. Why Bobby is against trading off I have no idea. He will, however, let us pay a hotwalker or exercise rider to feed for us. The going rate is a dollar per horse. Even that four dollars stretches my finances. I make a bargain with Charlie and Fourfoot. Three dollars plus one dozen home made cookies. The kids love it, even offering to buy more cookies. What they don't know is these cookies are from a mix. For two dollars I can make sixty cookies. I need to stretch my money as far as I can.

In this manner, I manage to have one afternoon off per week, sometimes two. **Dark days** (days when there was no racing) and Thursdays are best. Since we are paid Fridays, the kids are usually broke on Thursdays and more willing to feed then.

To stretch their paltry entertainment budget, Fourfoot and Charlie come up with an idea – the buck/buck parley. The minimum bet on the track is two dollars. They each contribute a dollar and peruse the racing form for a likely horse to 'hit the board'. Finding one they like, they place a **'show'** bet. If the horse finishes in the top three, they cash their bet and then bet it back (parlay) on the next race. If they cannot find a likely candidate in a race, they simply skip betting it. With any luck, they could well be betting for much of the afternoon before they lose their investment of a dollar each.

Sundays are more relaxed around the barn. Bobby buys day old donuts and invites us to help ourselves. The owners show up and visit with the grooms of their horses. Bobby is apt to launch into telling a story. Whenever he starts to talk, I, along with the others, mosey down the aisle to listen. Bobby always enjoys an audience and never minds that we are neglecting our work. Our work gets done in its own time. 1978 is the year of the Affirmed/Alydar duel in the Triple Crown. All of our barn help crowds into Bobby's tiny tack room to watch the three races on a miniscule television set. As Affirmed holds off Alydar in the Kentucky Derby and Preakness Stakes, the anticipation grows for the match up in the Belmont Stakes in June. For the third time, Affirmed triumphs over the other colt for the same one-two result and wins the Triple Crown. It is only the second time in my lifetime that a horse has won the Triple Crown, the pinnacle of horse racing. Bobby is effusive in his praise. It takes a truly great horse to win the triple Crown, an even better one to fend off such a terrific stretch run.

The track is a second home for Bobby; he virtually grew up here. He started race riding at age 14, when the rules were not as strict. He was a master of the half mile tracks that characterized Ontario tracks at that time. The lifestyle took its toll and Bobby, like so many of the older racetrackers, became an alcoholic. While he eventually recovered, the slight slur in his voice betrays his former life. The child jockey known for his fearlessness has now become a quiet fixture of Woodbine.

I learn that keeping my horses sound and happy is the best way to ensure my job is secure. One day, as I am doing up my horse, I reach for the Absorbine and realize it is missing. The last place I had it was in Tammy's stall, a few minutes ago. Panicked that I have left a glass bottle in an occupied stall, I quickly start tearing the deeply bedded stall up, looking for the missing bottle. I breathe a sigh of relief when I find the bottle intact, leaning against the wall. I shudder to think what would happen if the horse stepped on it and drove the glass up into his foot.

By now, Captain, the stallion I rub, has not stood up to training and is retired. Bobby accepts a two year old gelding named Major to fill the stall. Major has already been ruled off a track somewhere in the States because he is a 'bad actor' on the track and in the gate. Bobby decides to bank on his years of experience and try training the horse. Bobby gallops the gelding himself and starts to make progress. Then the horse's stall begins to develop a peculiar, obnoxious odour. I point this out to Bobby and my trainer decides the horse needs to be **physicked**.

This is an old fashioned remedy to 'clean out' a horse. Some of his manure is made into a slurry which the vet them puts directly into his stomach via a tube down the esophagus. For three days, the stall is so rank with the stinking diarrhea that I do not allow anyone into it. Even with my poor sense of smell, the stench is enough to knock me out. The crude treatment works and the colt returns to normal.

Shortly thereafter, Bobby gets the young horse **okayed** from the gate and enters Major in a race. The horse runs fine without distinguishing himself.

After the race Major will not cool out. For 3 days, the colt stays high on adrenaline.

After the race, Major will not cool out. Every time I put him back in his stall, he starts to **weave** at the **webbing** and kick the wall. After bathing him a second time to cool him down, I end up bandaging his legs to prevent him from injuring himself. He still does not settle. For three days, the colt stays high on the adrenaline and continues to weave and kick the boards. After he destroys several hard wood boards lining his stall, the vet is forced to tranquilize him. Within minutes, Major settles down and begins to eat quietly.

For the month of August, the entire **A meet** moves to Fort Erie, directly across the border from Buffalo, New York. This is one hundred miles from Woodbine. We pack up every bit of tack, equipment and all the horses. Not a screw eye is left behind.

In Fort Erie, the financially better off trainers and jockeys leave their wives behind to hook up with their girl friends. the Boyfriend and I decide to share digs down there as well. He finds us a cabin to rent for the month not far from the track. Since my financial resources are already stretched, I give up my apartment in Toronto to move in with him.

With the races in town, Fort Erie becomes party central. All the restaurants and pubs do a booming business for the month. There certainly isn't much else to do. Drink and screw and race, that is about it. I hate it. I hate the isolation of the town, the lack of mental stimulation. Since I do not drink and have no money anyway, I take a decided dislike to Fort Erie in particular and the month of August in general.

I do enjoy it when I hear the Boyfriend come home every day, whistling happily. It brightens up my day. With our busy work schedules, I do not see much of him until late afternoon, when we head out to eat at one of the local restaurants.

Surprisingly often, he arrives home in the early afternoon to announce, "We have an invitation to a party." There is something about his magnetic personality that people are drawn to. They want to be around and he is constantly being asked to join strangers for a meal. I watch him closely, trying to figure out how he encourages this popularity, as if this was some sort of magic dust that I hope will rub off on me. I have never developed the social skills most children learn in school and have only recently begun to make a few true friends.

By now, with a combination of constant exercise and preparing my own meals, the weight I gained while sick in college has easily slid off. The puffiness is gone from my face, I have dropped from size 16 to size 10 in pants.

One morning, Bobby comments to me, "Jan, you must have lost 25 pounds this year."

"Actually, it is 30 pounds," I reply. I am delighted that people are noticing how much thinner I have become. My private hope is to eventually gallop horses.

Bobby is sure that trainer Frank Passero, his nemesis, is going to claim a horse off us. Bobby strikes preemptively and takes one from him. In mid August, we run Pooh Bear at Fort Erie. I am standing down by the wire while the horses are on post parade when I hear the claim man bark out the dreaded "Number seven."

"Here," I reply weakly.

"Take him to the paddock." Pooh Bear has been claimed.

The horse wins by an easy three lengths, ears pricked. Bobby appears from the stands with one question. "Did we lose him?"

"Yeah."

'Who took him?"

"Passero." Frank has retaliated by claiming Bobby's own horse.

As I hold the horse for the win picture, there are tears in our eyes.

Credit: *Michael Burns*

Bobby appears from the stands. "Did we lose him?"

Bobby claims a sweet gelding named Delightful Tale to fill the stall. The horse is thin and looks run down. I notice in the **Form** the horse has been running poorly for a while. After the race, I poultice and bandage his legs and pack his feet with mud.

The following morning as I am picking the mud out of the crevasses of the foot, I strike a soft spot in the middle and draw blood. A horse's foot should *never* bleed. I quickly find Bobby to show him. Bobby, too, is appalled. Right then and there, he decides to give the anguished horse a full month off to recover his form. When the vet comes by, she puts the horse on antibiotics. As per instructions, I wash the feet out well and pack the crevasses with cotton soaked in Kopertox. This medication will kill the thrush fungus that has rotted the **frog** so badly it penetrated the underlying structures. No wonder the horse isn't running well and losing weight.

Every horse in the barn is now fit and racing. Only Tuesdays are dark (no racing) at Fort Erie and we run horses nearly every day. The strain of being at our barn up to eleven hours every day, six days in a row is showing. I am burnt out and cranky all the time. I have now worked seven months without a day off. On top of that, I cannot even get an afternoon off each week. In the third week of August, Shirley forbids me to take the dark afternoon off. "You had an afternoon off last week and the week before that. You do not need one off this week."

Credit: *Michael Burns*

The horse (Pass to Win) wins by an easy three lengths, ears pricked

Maggie is beginning to break down. Pooh Bear is gone. Delightful will not run for a month or more (but then he wins). Despite loving the horses, I am not cut out to spend all day at the track, with no other outlet for my creative interests. The Boyfriend is encouraging me to spend more time with him.

I finally give notice and quit. I never get the stake money owed for Pooh Bear's win. John Doe, now one of our grooms, has been keeping score. Since February, I am the forty seventh person to quit.

Chapter Six

NOT THE BEST

By the end of summer, when I arrive back in Toronto, I have no job and no place to live. I solve the problem of housing temporarily by phoning my elderly aunt who lives in downtown Toronto. She agrees to put me up until I can make other arrangements. She enjoys my company and even packs my lunch for me every day.

The Boyfriend finds me a job grooming with his friend Carl, making vague assurances that Carl will let me get on horses sometimes. I am so eager to ride I overlook my less than stellar impression of this trainer. A job is a job, when you have no money. Plus, a chance to gallop is a carrot I cannot ignore.

Carl is an Englishman who came out of the abusive apprenticeship system of the 'old country'. I soon learn this man is clearly incompetent but a shameless self promoter. Fortunately, he leaves the management of the stable to his able assistant trainer Teeter. Teeter is trying to improve his station in life but lacks the self promotion skills so necessary to obtain clients and become a trainer in his own right. Carl does not suffer from this lack of confidence.

Carl also belongs to that cadre of men who think all girls are here for men's enjoyment. I feel sorry for his overweight, unhappy wife who he drags to the track each weekend. She is probably well aware of his philandering and bears it only because she has so few options in life. I'm careful to keep a wary eye out and not be caught alone with this man. Fortunately, this is easy because Carl shows up late in the morning and leaves as soon as training is complete. Not the most dedicated of trainers.

Teeter assigns me four horses. I have only been on the job a few days when, while picking one particular filly's feet, I draw blood. Alarmed, I call Teeter (Carl, of course, is never around to tell). He tells me her previous groom was notorious for not picking the feet properly. This is the third horse that groom looked after that has developed severe thrush. Teeter also informs me the filly is entered in the sale a few weeks away.

By now, Teeter and I are in agreement on keeping Carl out of the loop. We decide to doctor the mare up and not to inform Carl. I dislike how this trainer chooses to impress his owners by insisting that the horse's hooves be oiled and look shiny while ignoring the rotting feet underneath.

After I pick the mare's feet clean, I press the Kopertox soaked cotton deep into her foot crevasses where the infection is. In days, the Kopertox kills the thrush fungus. Thrush, which is usually due to neglect, should never happen at the track when horses feet are picked twice daily.

Teeter shows me the filly's pedigree in the sales catalogue. She has a half brother who had earned one hundred thousand dollars, grinding them out the hard way, in claiming races. When the filly races a few days later, her form improves greatly over what she has shown all summer. The filly then sells cheaply and somebody gets a good brood mare at a bargain price.

I run all the horses in the barn. We have a gray mare entered in a minor **stakes** race. While I don't believe she will **get a ham sandwich**, I dutifully braid her up with the pink and white **colours** of her owner. I am dressed in a matching white shirt and pink vest and pants in which to run her. When I get to the paddock, the paddock judge mentions to Carl how smart the horse and I look. He remarks "If I hadn't chosen groom of the week yesterday, I would have given it to your girl."

Groom of the week is an attempt by the Jockey Club to encourage grooms to tidy up for the paddock. It comes with a twenty five dollar bonus, which is not to be sneezed at when my wages are only one hundred and twenty bucks per week. This strikes me as highly unfair that the contest is 'closed' already. After all, this is only Friday and the weekend is already excluded. It is the last time I bother to do the extra work to braid a horse up for a race.

The Boyfriend shows up occasionally to chat with Carl and check in on me. Despite asking him to put in a good word with Carl that might get me on horses, nothing ever comes of it. I had pinned my hopes on the Boyfriend getting me an opportunity to ride. I never even get a chance to shedrow a horse.

Both Teeter and I are getting increasingly frustrated at this job. Despite our best efforts, irritating incidents continue to hamper our work, often for no apparent reason. The boss, of course, is oblivious to any problems.

One morning, Teeter remarks to me, "What else can go wrong?" We are about to find out.

That afternoon I **paddock** a gray gelding named R J's Best. It is a typical claiming race in mid afternoon. We are running in the **four hole**. In the program earlier that day, I noticed the horse running in the three spot is **by** Bengal. The horse world is so very small. Just prior to starting at the track, I worked at the small farm where Bengal **stands** at stud. I am curious to see the son of this stallion. As we are saddling up R J, I look next to me where they are struggling to get the saddle on a chestnut, who is having none of it, and think, "That's not a Bengal."

There were three stallions on that small breeding farm. Bengal is notable for his placid temperament and his **dark bay** (almost black) colouring that he always thows. There is another stallion at that farm which is almost his total opposite. Spun Gold is a chestnut, much finer of build who always **throws** a screwball temperament.

The breeding book for that farm was just a calendar with the bookings scrawled onto it. The records were shoddy, at best. With blood typing still years away, there is no way to actually prove which stallion has bred your mare. This certainly wouldn't be the first time the wrong horse has covered a mare. I have seen three seasons' **get** of Bengal's and have never seen him throw a chestnut. The horse next to me is a chestnut, is lightly built and certainly has a screw loose. This horse keeps throwing himself and trying to flip over. I am certain he is by Spun Gold, not by Bengal. I follow the chestnut horse onto the track, leading my gray gelding. I hand the horse over to the pony boy and slip into the space by the finish line to watch the race.

The race starts in the usual manner. Until the quarter pole, the race is anybody's. Suddenly, there is a collective gasp from the crowd. Leaning over the rail, I can just see that two horses have gone down. There is a chestnut who's fallen and a gray horse that flipped over him. I have the only gray horse in the race. My stomach contracts in fear.

The moment the last horse passes under the finish line, I duck under the rail and start running frantically up the track. I look foolish running up the track but it is the only way to reach the injured horses. From a quarter mile away, I can see the gray horse has managed to scramble to his feet. Even from this distance, I can tell his right foreleg is broken because he is running back towards the barns on three legs. The reins must have flipped over his head when the jockey fell off because the reins are dangling against his chest.

The excitable chestnut gelding now lies still on the track, his neck broken. I pass him by with no more than a glance. Because R J is only on three legs, I manage to catch up to him by the three eighths pole. I grab the reins as he tries to pull loose.

The full horror of the accident now hits me. My poor horse. R J's foreleg has snapped off below the knee and is only hanging by the skin, yet he has managed to run an eighth of a mile on it. He is crazy with the pain. He alternately rears and flings the broken limb at me or attempts to stand on it, his weight sinking down onto the leg which cannot support him. It crumples hideously under him. The glossy white bone juts out from his upper leg, jagged where it is snapped off. Cherry red blood pours down the leg and mixes with sweat and sand. It is a sickening sight. The young jockey, who has already pulled his saddle off, looks on helplessly.

The track vet appears by my side but the horse ambulance has driven over to the other stricken horse. Some of the gate crew has come to help. For some reason, the **meat wagon** stays with the downed horse while the vet yells at the others to come to us first. Finally, they do come. A quick injection and the agonized horse drops to his knees and rolls over. He is dead instantly, mercifully.

As soon as the gray horse is down, someone whisks off the bridle and hands it to me, before rigor mortis sets in. I am too stunned to know what to do next. The carcass of the chestnut has already been winched into the horse trailer. To spare the onlooking crowd more of the awful scene, it is decided to pile the body of the second horse on top of the first.

As the men work to maneuver the second horse into place, somebody shouts, "Watch that sonofabitch don't kick you!"

Despite ourselves, the gallows humour breaks the tension and gets a belly laugh. Such is our way to deal with the double tragedy. Those of us who work on the backstretch develop thick hides to handle the heartbreak we see. Our pain never goes away; it just gets buried.

Still carrying the bridle, I trudge back towards the barn. The adrenalin that has carried me through the emergency has worn off. I feel hollow and sad and somewhat sick. I meet Teeter coming back from the test barn where he was looking for me when I didn't show up after the race. Mystified, he asks "Where's the horse?"

"RJ's dead," I reply dully. I am still numb with the shock. I briefly tell Teeter what happened.

I hang up the bridle in the tack room. There is nothing more to be done at the barn since Teeter has already fed the horses. I drive straight home, relieved to have quiet time to think. Next morning, I hand in my notice. I have worked there just four weeks. Teeter quits soon after.

The job has proved to be the worst one ever, for me. I am seeing how much a job is a reflexion of its boss. Carl spent too much of his time promoting himself, trying to make himself look competent for his ignorant owners but failing to provide an adequate environment for the animals. He is a lightning rod for trouble.

I am just beginning to learn to trust my judgment and intuition about people. I still have a lot to learn.

Chapter Seven

AURORA MEADOWS

Having spent the summer grooming and walking horses, I am desperate to get a job galloping. Someone advises me that breaking **yearlings** is the best way of getting the necessary experience for snagging that job. Jimmy, the resource person who originally placed me at Woodbine, finds me a farm that is looking for girls to break yearlings. I phone the farm and am invited to come up and ride for an interview.

The forty mile drive through the rolling hills of Ontario has some of the prettiest vistas in the country. The roads are lined with mature trees, fancy estates with their sprawling mansions and neatly painted paddocks of horse farms. When I arrive at the farm, I am met by Fran, my prospective new boss. With no preamble, he produces a set of tack which he drops in front of the farm pony's stall. "I'll be back in fifteen minutes" he states as a matter of fact. I am being tested on basic knowledge.

I quickly knock off the pony and saddle him up. I am just putting the final keepers on the bridle when I notice the running martingale lying on the floor. I silently swear to myself. This should have been the first piece of tack on the horse. I remove the bridle to slip the martingale over the pony's neck. It needs to be let out to accommodate his thick neck. I have to redo girth, centering the martingale between his front legs. I pass the reins through the martingale, taking care not to thread the reins backwards. I am just finishing the bridle when Fran reappears.

He quickly checks the tack for adjustment without comment. He leads me to the arena where I tighten my girth and mount. I only ride for a few minutes, demonstrating my posting trot and canter on command. When I pull up, Fran has a further request.

"I want you to just lay over a young colt." Fran leads me to another stall where he brushes off a yearling and puts a set of tack on him. I follow Fran down to the arena again. While I grasp the reins in my left hand, Fran boosts me up until I lay across the saddle. This young colt has never been ridden! We repeat this manoeuver a few times, and then Fran helps me find the stirrup with my left foot. I slowly slide my right leg across and sit up, then cautiously reached for the other stirrup. While Fran grasped the colt by the chin strap, I gently try out the steering. So far, so good.

Within minutes, I am turned loose and am trotting around the arena. I have just **backed** a horse on my job interview and with that, I am hired.

My co-worker is a terrific rider named Gwen. She trained in England as a riding instructor for two years. While I find her a bit frosty, we work so well together that we give ourselves the name, "The Efficiency Team".

The farm is conveniently laid out. All the barns are grouped together in the front half of the property. There are numerous paddocks close to the barns. A long lane way leads to the back of the property. Off this are larger paddocks, used for broodmares and the last one reserved for riding in. Fran runs a tight ship, as I am to learn. There are only six full time staff, plus weekend and part time riding help.

I am living with the Boyfriend now. His ex-wife moved into his apartment while we were in Fort Erie and is still there when the meet at Fort Erie ends. One day, the Boyfriend phones me to meet him; I assume he is calling when she is out. When she opens the door to me, I realize I am mistaken. The boyfriend swears at me but

I truly thought he said to meet him there. The result is his ex-wife moves out and I naively move in. A buddy of his rents the spare bedroom.

The apartment is located on the edge of the city and while the weather is fine, I drive to the farm each day. The work day at the farm starts at 7 am. We ride five sets in the morning with a quick coffee break after the third. At first, there are four of us riding. Fran, Gwen, myself and a local girl who only rides as needed. We give our yearlings a quick groom and throw the tack on as fast as possible. The four of us ride a mile down the lane to the back field. There, we split up and trot on our own, teaching our youngsters to turn and tolerate other animals coming toward and away from us. Only in the last week of their training do we pair up to gallop.

After the riding, there are also the not-yet-backed horses to be **lunged** or driven. Fran outlines a six-week breaking program that works well for him. On Monday, the yearlings are tacked up with bridle complete with rubber D bit, exercise saddle and attached **side reins**. The young horses are left in the stall to accept this new equipment for an hour or two. Then they are put out in a pen to allow them to move about. By Wednesday, we put a lunge line on them and, using the round pen, teach them the basics of lunging. This continues all week, with Sunday off.

The following week, ground driving begins. Ground driving is fun. The first time a colt or filly feels the lines on them, they may panic and run, so we have someone at their head until the horse begins to relax. We begin driving in the arena but soon slide the door open and drive around the entire barn area. Gwen mentions weaving around the trees planted in the driveway so I try that too. At the end of the hour, we drive the youngsters down the aisle and right into their stalls, where we carefully pull the lunge lines off of them. This year we have a abundance of chestnut colts with white **blazes** and **socks**. This leads to a funny incident. Gwen is driving a colt one Wednesday when Fran stops her to ask her how this particular colt is going.

"Best one. I just walked into the stall, hooked him up and drove him on out," Gwen replies.

"That's good," is Fran's comeback. "He's never had the driving lines on before." Gwen mistook the colt for one in his second week of breaking, not in his first. She could have been badly kicked. Instead, the good natured colt had his breaking process shortened by a week.

On the third Monday, we back the young horses. After the two weeks of ground work, most of the youngsters take to riding like ducks to water. Within minutes of laying across them, we are trotting around the arena. Few of them ever buck. One trouble maker stands out from the rest. She's a tall, skinny filly who sticks her head out between the fence boards when she is turned out on her first day. She gets caught up between the rails and in her panic, breaks the fence apart.

All of us on the farm set out to catch her. We finally corner her behind a gate where there is a short six foot section of fence. I cannot believe my eyes when she simply plows through the solid three plank fence, snapping the rails as if they were matchsticks. As she starts down the lane way that leads to the back field, one of the workers picks me up in the pick up truck to chase her. We are gaining on her just as she turns into the back field. As she gallops across the field, the driver accelerates to cut her off. Concentrating on the filly, we completely forget about the slough in the middle of the field. With the truck accelerating, we drive straight into it. Now the truck is mired up to the axles.

We finally manage to catch the tired filly. The tractor is dispatched to pull the truck out of the mud. The farm returns to normal and I have learned that solid fences will not necessarily hold a horse determined to escape.

We break forty babies that season. At this point, every yearling that Fran has broken has gone on to break its maiden. Less than half of the race horses make it to the races, much less break their maiden. This has been a great education for me on how to start horses safely and efficiently.

I eat lunch by myself in the coffee room. When I learn that the farm has a direct phone line to Toronto, I make a practice of phoning the Boyfriend at lunch. His buddy always seems to be around when I am home and I do not get any other chance at a private conversation. These intimate conversations smooth out some of the difficulties in our growing relationship.

After lunch, Gwen and I bed down all the stalls, knock off horses that were ridden in the morning and help bring in the horses. There are nine broodmares in a field half a mile down the lane. It is time consuming to make multiple trips to bring them in. Soon, Gwen and I figure out how we can bring all nine in, in one trip.

We take two shanks apiece. One shank is clipped to the far horse, then doubled through the halter of another mare and back to our hand. Then we catch another horse with the other shank. When we each have our three mares, we open the gate and let the remaining mares follow us. Once we get to the barn, we catch these other mares and sort every horse into the correct stall. Since these are pregnant mares, none of them run around like idiots and no mare is ever hurt. There is only one fly in the ointment. Laudnaer is the easiest to catch but a pig to lead. She always drags behind the others. With three horses in hand, she is like pulling a dead weight. I always hope Gwen catches Laudnaer first and I do not have to deal with her.

Fran knows what we are doing, turning a blind eye because everything is working out so far. We assure him the mares will be fine – which they are.

We have quite a number of weanlings on the farm. One of them, a good looking bay, just does not want to be caught. He does not run away, just stays out of our hand's reach. We try everything. He is turned out just with the pony, which he will docilely follow into the barn, where he is shooed into his own stall. Regardless, he remains difficult to catch even when we try handling the colt in his stall every day to make him friendlier.

One day, at coffee break, we are discussing this. Eventually, we turn this youngster out with a soft cotton lead attached to his halter and let him wander about with this. At first, this works well. Our day is considerably easier not having to trick the colt into being close enough to catch. Inevitably, the worst happens. The colt sticks his head between the fence rails, gets the lead line stuck and pulls his halter off. That afternoon, all six of us are in the field long after the other horses are in and fed trying to catch this nervous animal. Over and over again, we quietly walk about the colt, trying to corner him. He will let us get within a couple of feet before evading us. Finally, one of the guys corners him, makes a leap and grabs him by the ear. We quickly halter him and he quietly follows us in. We are a full half hour past quitting time. I left for work at 6 am and it is after 6 pm before I return home.

As if that wasn't bad enough, the buddy of the Boyfriend makes a pass at me while the Boyfriend is out. I am furious, demanding the buddy be kicked out and reluctantly, the Boyfriend does so. We have a heart-to-heart discussion about not sharing the apartment with anyone. Now I believe my home life will be easier.

The very next day, a friend of his pops by. She announces that it would be a great idea if the three of us share an apartment. When the Boyfriend enthusiastically agrees, I lose my cool. Eventually I simmer down enough and somehow they talk me into looking at a bigger apartment. I am very relieved when the deal falls through.

With such a long drive to the farm, I have developed the habit of sliding into work about ten minutes late. One morning, I am twenty minutes behind. The others are already prepared to ride.

Fran lays down the law. Enough is enough. If I am late once more, I will be fired. I get my butt in gear and started showing up on time. Fran is tough, but fair.

One of Fran's responsibilities as manager is to make the farm profitable. As racing closes down, he takes in more racehorses for layup to be conditioned for the start of racing. Even the arena is turned into stabling. Twenty temporary stalls are erected in the arena with an aisle down the centre. There is still a track wide enough to ride around the stalls. With the riding finished for the autumn season, Gwen and I have enough time to care for the extra horses.

As winter approaches, the roads may become impassable so I move into the farm houses at the end of December. With no one to share the cost, the Boyfriend decides to give up his apartment and move back onto the track.

New Years Eve is unusually mild and it pours rain. The Boyfriend and I have gone over to the home of some friends of his to spend the evening playing cribbage. As I survey my astonishing hand of cards, I try to figure what move will gain us the most points. I have three fives and two jacks.

"Lead with anything," the Boyfriend instructs.

When I lay down a five, he suddenly turns on me and erupts in anger. Trying to defend myself, I retort, "You said to lead with anything!"

This quickly degenerates into a fight and suddenly, the Boyfriend disappears into the night. I am embarrassed and angry. When I calm down, pacified by my hostess, I remember the boyfriend has my car keys in his pocket.

The hostess bundles me into her car and we go looking for the Boyfriend. When we cannot find him, I realize he has power-walked back to the track. I have the woman drop me off by the deserted back gate at the track. I squeeze under the fence and find the Boyfriend in his tack room. The walk has cooled him down and I share the narrow bed with him for the night. I never seem to know what will set the Boyfriend off to avoid these public scenes. In the morning, he drives me back to fetch my car.

Overnight the temperature has plummeted and the flash freeze turns everything into solid ice. At the farm, all the horses must stand in their stalls for the next two weeks. There is no chance to turn them loose. Seven of them begin **cribbing.** All we can do is give them lots of hay. The exceptions are the horses in the arena. When we turn the entire group out, they gallop happily around and around their stalls until the whole group is exhausted and steaming.

There are no horses in training the first two weeks of January. On the second of January, Fran reaches into his desk and extracts a couple of **pulling combs**. He presents them to Gwen and me, saying "There are 100 horses on the farm. You have two weeks to pull all their manes."

Manes are thinned and shortened by pulling out the longest hairs. While this sounds barbaric, horses have few nerve endings in their mane and most stand still for the grooming task. Long and thick manes get tangled in reins, causing a hazard for galloping.

We each have to pull an average of five manes a day. We leave the most difficult horses that need to be held until last. My fingers become sore but my ability to pull manes improves in a hurry.

After two weeks, the ice thaws. Finally we can turn the horses out. The weanlings, which are now officially yearlings, are exuberant. As we turn that herd loose, we count them as they fall in their excitement. There is no way to prevent these youngsters from running on ice. They just have to learn. One of them punctures his knee and has to go back on stall rest while it is being treated. None of the older horses make that mistake.

With the racehorses back into training, Fran needs more riding help. He hires a girl I know from the Humber College Equine Program from the class behind me. On the morning she starts, another girl from the same class shows up for an interview. As we greet one another, Fran asks how we know each other and I explain how we were all at Humber together.

"I've never had any luck with Humber students before" he remarks. I know my work has changed his opinion of Humber and I am not worried for the other two girls. They both work out well.

Fran prefers to hire girls with pleasure riding experience over seasoned exercise riders. He keeps threatening us with making our horses jump. I assume this is an empty promise. Then one morning, he pulls down some straw bales and lays them in a line across the arena.

None of these horses are broke to jump but we have fun with the exercise. To my surprise, some of the best jumpers are mediocre race horses. Some of the horses cannot seem to get the hang of it, leaping awkwardly over the straw. None of them are the worse for wear by the end of the day. Horses and riders both benefit from the change of pace.

Fran breaks up the monotony of riding indoors by sometimes taking us across the road to ride in a field. It can be bitterly cold. One morning even Fran, wearing his skidoo suit, complains in the minus 40 wind chill. I have no money to buy ski pants and only have long johns under my jeans and must grit my teeth to deal with the cold.

Gwen and I handle all the breeding stallions as part of our duties. Many farms do not allow girls to handle stallions, fearing the stallions may attack the women. Good Port, one of the breeding stallions, is so easy to handle that we lead him out to his paddock with the shank just clipped to the bottom of his halter. The other stallions have to be handled with a chain across their nose.

About this time, the farm leases another stallion for breeding. Fran warns us about the new horse. "Don't take your eyes off him for a second. He has a mean streak in him."

The legendary blacksmith Ernie Jay is the only farrier allowed to trim him, coming to the farm especially to do the stallion. I learn from the blacksmith that this horse has foundered so badly the coffin bone had dropped through the sole of his foot. With the constant pain of his feet, the horse has turned mean. I am grooming him one afternoon when he leans back on his tie chain and stretches himself. I watch him casually. Suddenly he leaps for my arm with his mouth wide open and grabs me through my puffy winter coat. Despite my caution, he manages to break the skin. I still carry a small scar.

With a hundred horses on the farm, Gwen and I are busy. By noon, the part time help have gone home. Gwen and I still have to bed all the stalls, knock off the riding horses and bring in the breeding stock.

One morning, as Gwen and I walk into the arena with our first set, we find the old man who does the feeding collapsed on the floor. He has been lifting a heavy pail of oats from the bin when his back gives out. He cannot even move without assistance and has to go on disability. On a small farm like this one, the loss of even a single worker creates more work for the rest of us.

A new filly arrives for breaking near the end of February. She is a big two year old, dark gray and good looking. Because she is virtually unhandled, I am tying her up with binder twine instead of the tie chains we use on the older horses. Binder twine should break more easily instead of the halter if the filly pulls back. I have switched to using the plastic twine because it is stronger and does not wear out as quickly. As I pull the loop with my index finger to form the slip knot, the filly throws herself backward. Instantly, I realize my finger is trapped and the only hope for my finger is if the binder twine breaks, since the panicked filly is too far away to reach. For a few awful seconds, I pray as the cord bites through my skin. Finally the twine breaks.

As I look at the blood streaming from my hand, my first aid training kicks in. I clamp my other hand firmly over the wound. I let myself out of the stall and carefully close it before walking into the office/coffee room where I know Fran is working. I uncover the hand torn down to the tendons to show Fran and calmly state, "I have to go to the hospital."

Fran turns three shades of green. Obviously he has no stomach for blood. Luckily, the farm owner is sitting in the office with Fran and offers to drive me. The whole way, he keeps assuring me, "Take it easy. Take it easy. Take it easy." His driving is so erratic that after a while, I start to reassure him, "Take it easy. Take it easy." Like so often with a major injury, the pain does not really kick in for about 15 minutes. By then, we are close to the hospital. By the time we arrive, I am begging for pain killers. The attending doctor tells me I am lucky. The twine sliced to the bone, then along it without touching the tendons. He finds my skin so toughened from farm work that it is difficult to pierce with the needle as he stitches my finger. This turns out to be a blessing. I have since heard of three other people doing the same thing, having a horse jump back as they were tying it with twine. All three lost the end of their finger. I realize how lucky I am.

Some months after this, when the filly has been successfully started and is moved to the track, she flips over completely, hitting her skull hard. For three days, she lays on the floor and never recovers. Finally, she is put down. I wish I had been in a better position when I was grooming her and could have taught her not to pull back when tied.

With stitches in my hand, I cannot work. I go to my parents to recover. Now there are two injured people laid off, the old man and I. After several days, I get a phone call from Fran. They are struggling along short handed. Can I come back and help in any capacity?

There is no way I can ride since I cannot close my hands on the reins. I can however, hold a curry comb and hoof pick. Grooming proves to be the best rehabilitation possible for my hand. I brush off the horses in training allowing the riders to simply throw their tack on. When I finish those, I start the afternoon knock offs. One day, I rub over forty horses if you count the ones I do twice.

Ten days later, the horses return to the track and I am laid off.

I am somewhat peeved at being called back in only to be discarded so callously. The truth is, this farm runs a lean operation. I have gained a lot of skill here but I still have a lot to learn about being a professional.

Chapter Eight

MR ANNESLEY

"How can a horse look that good and run that bad?" Gordon Annesley asks of no one in particular.

It is the day following Romantic Air's first start of his four year old season and the horse has run **up the track**. Again. He is still a maiden. The ebony gelding is gleaming from the grooming I am giving him but there is a race track saying 'there's no such thing as a good sound horse'. If a horse has any heart, he will eventually develop the problems that plague good running horses - '**ankles**', '**knees**', **bucked shins**, **bowed tendons**. There isn't a pimple on this horse. Romantic simply hasn't tried.

I started with Gordon in early March. Because I have not yet recovered from the injury to my finger, I take a job as a groom. But I ache to gallop. I am burned out by the long hours of the previous year and find grooming irksome with its tedious demands. Some people are happy to spend all day working on their horses. Not me. I want to have a life outside of the track. However, Gordon is so kind that I settle in...sort of.

Now that the Boyfriend has moved back into a tack room on the track, I have no one to share an apartment with and cannot afford to rent one on my own. Neither is staying with my aunt a viable option long term. With no other choice, I move into the girls' dormitory at the track, located just inside the backstretch gate.

Living there is even worse than I feared. The guards at the entrance to the backstretch keep us quite isolated, not even notifying us if we have a visitor. No men are allowed anywhere near the dorm and there are no telephones except a pay phone outside at the end of the building It feels more like a prison than a home and I immediately have nightmares about being in jail.

When full, two girls share a room. Each room has only two beds and dressers and a shared bathroom. I find myself co-habiting with stranger who does not always respect my belongings.

There are no cooking facilities either. The expense of eating every meal out makes the living arrangement more costly than sharing an apartment. I am very frustrated by my living arrangements. The Boyfriend, his pal Donny and I eat out seven nights a week. Then the guys play pool while I watch. With the girls not allowed on the backstretch after 6 pm and the men not allowed at the dormitory at all, there is no place where I can just hang out with the Boyfriend in the evenings. I start spending some nights in the Boyfriend's tack room. I leave my car at the dorm and walk to my barn at feed time. This fools the guards as to where I am actually sleeping. Other times, the Boyfriend sneaks me in to the backstretch, hiding me in the trunk of his car. It is pretty silly that women cannot live on the backstretch. This is 1979, after all.

I use the time management lessons I perfected the previous year to zip through my chores to be out of the barn by noon. I hate to get hung up, waiting for a horse to cool out before I can do it up, then be late leaving the barn to go home. If Gordon is watching a horse train and the pony boy shows up, I send a horse out. Poor Gordon comes back to find one horse still walking, three empty stalls not yet mucked and two horses back shortly. While Gordon is stressed, I put the one horse away, get the galloped horse out walking and am prepared to walk the ponied horse until the hotwalker becomes available. Finally, Gordon lays down the law. There is to be a maximum of two horses out at once. Now his nerves can recover.

At least, my work situation is better than last year at Bobby Fisher's. I am the only groom in my barn. My trainer Gordon is a quiet man who has grown up around racehorses. Both his parents trained racehorses at a time when it was unheard of for women to be working with horses. A true gentleman, Gordon is courteous to everyone. He is patient, thoughtful and unhurried. He gives his horses the time they need to develop and get fit. We have seven horses in training but only one winner. The rest are still maidens. When we run our only winner at Greenwood, the horse is claimed off us, leaving us with a barn full of maidens.

Of the seven horses in our barn, I groom four while Gordon himself grooms the other three. I feed six afternoons a week and Gordon comes in weekly to feed and let me have an afternoon off. Initially we are using a freelance exercise boy and have hired a hotwalker. Since I paddock all the horses, I am fortunate that we do not run all that often.

One of Gordon's owners is a lovely man named Mr. Surette. If anything, he is even more courtly and unassuming than Gordon. I rub all three of his horses – Romantic Air (four years old), Golden Javelin (three year old) and Gray Rap (two). Each Sunday, Mr. Surette comes to see his horses train and bring carrots for them. I have to hold the horses while he treats them. I cannot believe how delicately each of the horses nibbles the carrots off his proffered hand. Most thoroughbreds are not that mannerly.

Gordon does not believe in letting his grooms give nicknames to the horses in their care, believing that nicknames are in poor taste. Therefore, the horses in this barn are referred to by their sex and colour – The Black horse, The Chestnut mare, The Gray filly.

Romantic, the Black horse, is usually one of those horses **ponied** but after that first disastrous race, Gordon decides to find a strong boy to gallop him. At the farm, the black horse is pleasant to ride but at the track, he pulls so badly that most riders cannot manage him. Our freelance boy, Tommy, takes him out to the sand ring. On that eighth mile oval, the horse only pulls toward the out gate but lets up on the other side, so the exercise rider manages him fine.

One day, only a few days before his next race, Gordon takes the black gelding to the training track instead. When Gordon returns to the barn, he is so completely upset he can barely speak. The horse ran away on Tommy. Gordon blames himself.

"Well, he was all race horse out there," Gordon consoles himself.

Gordon walks him the next day and we cross our fingers that the horse is fine.

It is a slushy spring day at Greenwood when Romantic races the next time. What a different situation from the last race. Romantic wins with authority, a full two lengths in front of his nearest competitor.

I stand with the other grooms waiting for my horse to pull up and gallop back after the race. I wait. And wait. All the other horses come back and are collected by their grooms. I start to worry. What has happened to him? I anxiously peer down the track. Finally, I see the jockey Normie coming from the other direction. He never got the horse pulled up. The horse simply continued around the track for another lap.

Gordon's bandages are soaked and have slipped down Romantic's legs. One of them is missing altogether. The horse is still so strong that I have to push hard on the reins to steady him for the win picture.

The repetitiveness of grooming continues to grate on me. One day, I am in such a foul mood that I think to myself, if Gordon says one more thing to me today, I am going to quit. A few minutes later, Gordon does say one thing to me. "How would you like the afternoon off?" What is there not to like about the man? I come back the next morning refreshed and eager to do my job again.

Besides time off, there are other changes at the track. There is a new development in leg wraps. The manufacturers have introduced stretchy bandages with velcro closures. The change is universally adopted and soon bandage pins are relegated to history.

Credit: *Michael Burns*

Finally, I see the jockey Normie coming from the other direction. He never got the horse pulled up. Norm Davies up.

The new bandages are much easier to use and safer, too. Besides, they come in a wider range of colours.

There is very little bandaging done in this barn. Gordon believes in giving horses the time to mature and if a horse is not standing up to training, he sends the horse home for a rest. We have the soundest racehorses I have ever seen. When I see the bandages slip on Romantic in his race, I realize that Gordon actually does not have much practice putting bandages on.

Gordon is a minor trainer at Woodbine, relegated to the yearling barns at the south side of the track. These barns were only built to house sale horses for two weeks out of the year and never been intended for year round use. The stalls run across the barn, eight to a section and there is no shedrow around the outside. Instead, the barns are open to the elements.

I come in to feed one rainy afternoon to find the barns flooded. All of my stalls are saturated. One filly is sloshing around with water over her fetlocks, another horse is in two inches of water. Fortunately, the grain is piled on pallets in the feed room and out of danger of being ruined.

Another groom in the barn finds me a couple of empty stalls and I move these two horses for the night. I bed them down and bring them their hay, water and feed pails. Next I dig trenches in the sand to funnel water away from our barn. In half an hour, the water level in the barn begins to drop. I pick out the rest of the stalls, adding fresh straw where needed so that all the horses have dry stalls. What normally takes half an hour for afternoon chores has stretched to one and a half hours.

I decide to call Gordon to let him know where I moved the two horses. Since it is a race day (this is before cell phones), all the phones at the track are locked down until after the last race has run. I drive off the track to find a pay phone. Gordon immediately returns to the track and strips the two flooded stalls. He limes them well and freshly beds them before moving the two horses back into them. Now we can start the next morning organized.

My sister Anna is attending college and cannot find a summer job. She has put in an application for summer work at General Motors but has not heard back from them. Meanwhile, we are short a hotwalker. I ask my sister if she would take on hot walking for the summer. She is pleased to have some source of income and says yes.

When I approach Gordon about hiring her, he is pleased. "You tell your sister to come see me."

Anna and I are able to share a room in the girls' dormitory. I am relieved to move out from my former roommate. While there are no cooking facilities, a bar fridge and kettle are allowed. Anna has a fridge and we hide an electric fry pan under the bed and begin to secretly cook for ourselves. Onions are a no-no as the smell will give away the cooking. Not only do I save money but do not have to go off the track every evening. I spend some time hanging out with Anna instead of with the Boyfriend.

Anna learns to hot walk easily since she had often handled riding horses. The Black horse is the trickiest because he has to be walked with a chain in his mouth since he is so bullish to handle. To water him, the chain is slipped out of his mouth and he is allowed three swallows of water. With each gulp, he dives deeper into the pail. By the third gulp, his muzzle is in the bottom of the pail and the water up to his eyeballs. She has to then wrestle him out of the water and slip the chain back between his teeth.

Being in college, Anna has no money for proper footwear. I was not happy with her wearing running shoes but it is all she can afford. Because all the barns are open, whenever there is a loose horse, the animal stirs up all the horses being cooled out, making it dangerous to lead your own animal. The Gray filly in particular is very spooky.

The Boyfriend has picked up a job breaking a couple of two year old fillies on a farm about twenty minutes north of Woodbine. I go with him one afternoon to watch. A couple of days later, he is offered a prestigious job, through the buddy he kicked out of his apartment a few months before. The job is based in New York, galloping for one of the top trainers on the continent. The Boyfriend decides to take the American position. While I do not trust his buddy, there is nothing I can do.

With the Boyfriend leaving, the two fillies are barely started under saddle. I see an opportunity and hatch a plan. I approach the owner Henry and offer to break the yearlings for free. Although a little dubious, Henry agrees. I ask Donny, another pal of the Boyfriend, to help (I need a ground man) and I take over the breaking. I have enough time between morning chores and afternoon feed to get both horses done. The big filly, not yet named, is a swaybacked chestnut who is incredibly kind and easy to ride. The biggest difficulty is adjusting to her sway back. Snoopy, on the other hand, is like riding a greased pig. She eventually will drop everyone who gets on her. The day I watched the Boyfriend ride, she popped him off. Snoopy will drop me seven times, three of them just getting on.

Henry's daughter Rhonda is just a few years older that I. She joins Donny most days to watch while I ride. Soon, the three of us become easy friends. I ride each filly for about twenty minutes, teaching them to steer and trot or canter as needed. I start riding each filly in the arena which is attached to the barn. After a couple of weeks, Rhonda suggests riding them in the field next to the arena. It is a pleasant place to ride and the field is a good size for this work.

As I am finishing up my morning work one day, Gordon informs my sister and me that we will be claiming a horse this afternoon. With the pressure for stalls, if you lose a horse (due to being claimed or going lame) and do not fill the stall quickly, it will be reassigned to another trainer. The Jockey Club tries to keep its races full by keeping fit racehorses in the stalls. I am not pleased. Now I must tell Gordon I planned to be away from the track breaking horses in my free time. To allay my hissy fit, Gordon invites my sister and me to lunch at a family restaurant chain. We have a lovely time and it takes the edge of my annoyance. Of course, we are still planning to claim a horse.

When the time comes for claiming the horse, I wait until the race horses are well on their way to the paddock before I walk over with my halter and shank. I do not enter the paddock until the horses are in the post parade.

The paddock judge hails me. There has been more than one claim put in for this horse and there will be a draw for who gets it. We are each assigned a number on a ball and one ball is poured from the cup. It is mine. After the race, I stand by the numbered stall corresponding to the race number of the horse. The groom brings the horse in, faces it to the back and pulls the bridle. I slip the halter on and put the chain over the nose. The gelding is quiet and did not run particularly well. I bath him, my sister cools him out and I hang his feed tub.

A minute or two later, the groom next to my stalls says, "Jan, your horse isn't eating."

Sure enough, the horse barely touchs his grain all night. Perhaps the horse is the nervous type but with Gordon rubbing this horse, the animal soon settles and begins to eat better. As his appetite improves, this horse perks up and he becomes a healthy addition to our barn.

There is a perception in the horse industry that the racehorses at the track are universally badly treated. I actually find the standard of care far higher than anywhere else I work. The grooms spend the better part of their day looking after just four horses. Over the years, the grooms become remarkably skilled at handling these young animals and bringing them to the peak of their condition.

There are always exceptions. Farther down our barn, Leroy has a few horses stabled. Leroy is a loud, arrogant man and is universally despised. One of the horses he trains is a big, ill-mannered colt. The owner refuses to geld him, despite being told how dangerous the colt is. One morning, Leroy loses his temper and beats the colt about his balls with a chain, then sends the exercise rider out to gallop him.

Everyone in the barn is aware of the cruelty and is furious. Shortly after, we hear how the colt bolted through the inside rail, injuring the exercise rider. One of the older hotwalkers remarks to me, "You know, Jan, God does not always work that fast."

I start going down to my parents home a couple of times a week to telephone the Boyfriend. He is galloping for one of the best trainers in the States. I rack up an impressive telephone bill as we become closer than we ever have, now that there is no one to come between us. Now I can say what is on my mind and he hears me out. We make plans for our future. Maybe I can join the Boyfriend there.

At the farm where I am breaking the fillies in the afternoons, the Big Filly is coming along fine. Snoopy is more difficult. The problem with Snoopy is that although Donny is holding her while he legs me up, she can run backwards while I am in the air and I land in front of the saddle on her neck. There is no chance to save myself and always fall off. I don't know how to fix this problem.

One day, Rhonda's husband visits the farm. He is a trainer who exercises his own horses. When I tell him my problem, he offers to help me the next day. He starts by putting Snoopy in a stall. As he legs me up, she slams backward into the stall wall before I even find my stirrups. I kick her and she leaps forward. I pull her head around to prevent her bucking me off. For a couple of minutes, we go round and round the stall until I firmly establish who is boss. She never pulls that stunt of dropping someone as they get on her again.

Both fillies are riding reliably now. As the fillies begin to settle under tack, Rhonda offers to teach me the basics of galloping. She shows me how to cross my reins and take a double hold of both reins (called a bridge), then plant my hands into the base of the neck. I then straighten my elbows and lean over my hands. I am practicing this one day on Snoopy as Rhonda is watching (Donny is unable to come this day).

"I am just going to step into the barn to touch up her stall. Will you be alright for a couple of minutes?" Rhonda asks. I assure her I am fine cantering around the field.

No sooner than she disappears around the corner of the barn, Snoopy drops her head between her knees and throws herself into reverse. My hands slide down her neck and I hook each thumb under the bridle. It happens so fast that I do a somersault and land on my feet, holding the bridle. Snoopy jumps sideways to sprawl onto her belly. As I leap to try to catch her by the forelock, the filly is even quicker to her feet and runs loose. I call Rhonda for help catching the filly and to be legged up again. This is fall number seven from Snoopy. I am embarrassed and wonder if I will be successful at breaking her.

At the end of June, my sister Anna gets her call from General Motors. Of course she takes the job since it is easy work and twice the pay. She gives Gordon a week's notice. Three days later, Gray Rap spooks and leaps straight onto my sister's foot, breaking Anna's toe. Now we have a hotwalker that cannot walk. We trade jobs. I take over the hot walking and she tacks up my horses and mucks my stalls into the aisle. Between walking hots, I grab my pitchfork and carry the manure down to the dumpsters. I also have to do all the bandaging. Of course, I still have to do all the afternoon feeding.

To rub it in, the first thing my sister has to do on her new job is purchase safety steel toe boots. After Anna leaves, there is the perennial shortage of help to deal with. We just cannot find another hotwalker. On weekends, Gordon's brother comes in and give us a hand but that does not solve our issue for the rest of the week. The solution lies with Gordon's fourteen year old son, Chris. I am tasked with keeping an eye on Chris as he learns. He is a pleasant kid who tries hard and soon masters the knack of walking hots. There are a few times when I have to rescue Chris as he goes sliding by on the end of the shank, being dragged along by Romantic.

I continue to drive down to my parent's home a couple of times a week to call the Boyfriend in New York. I can feel it somehow when he wants me to call. I haven't talked to him in days, due to the pressure of work. Now I am restless, feeling something is wrong. I decide I will drive down late tonight. Because we raced today, I am riding the fillies in the early evening. Since Henry is home, I do not take Donny with me and can leave directly from the farm. The two fillies are finished their breaking and Henry has stalls at the track for them. I have just ridden the big filly and pulled up to a walk when Henry asks me to gallop her once more around the field.

I do as I am bid. As I am passing by Henry, I glance down and see the Boyfriend squatting in the grass beside Henry, a secretive smile on his face. I pull the filly up and stare open mouthed, unwilling to believe my eyes. The Boyfriend was homesick for me and has returned, walking out on his job.

It is awkward greeting each other with Henry standing there. We begin catching up but keep a modest demeanour.

I show him my new helmet cover. The Boyfriend bought an identical one while he was away. We marvel at how alike our minds have become.

That night, he checks us into a motel room up the road. It is good to have some intimate time with him.

The fillies ship into the track and my riding is finished for the moment.

Now that the Boyfriend has returned from New York, he writes a horse's name on a piece of paper. He shows it to all and sundry, declaring, "This horse will win the Kentucky Derby next year."

It is a filly and her name is Genuine Risk. He has been galloping her in New York and firmly believes in the filly's talent.

By now, I am sick of grooming. The Boyfriend is going to Fort Erie for the August meet, where he will be galloping horses for a small barn. He dangles the prospect of getting on horses at the track in front of me. I decide to go, too. I give Gordon notice. I actually choke up when quitting Gordon. If he had put me on a few horses, I might have stayed there. He was such a wonderful man to work for.

I am pleased about my future prospects. My relationship with the Boyfriend is solid, I have gained confidence in my riding skills (despite Snoopy) and I believe success is very close for me.

Chapter Nine

SUMMER OF DISPAIR

When the Boyfriend goes down to Fort Erie before I do, he rents a hotel room for us to share. I am disappointed in it. The hotel is pretty decrepit and there are no cooking facilities. It is late in the season and there is not much of a selection.

I settle into the hot walking. We are located in a small barn near the training track across the road from the main track and barn areas. It is a quiet backwater without much oversight by the officials. I have brought my cat down and settle him into the barn.

At first, since the Boyfriend has only been home a couple of weeks, we are closer than ever. We spend most of our free time together although we are always visiting other people. The Boyfriend always seems to have a wide circle of friends and we spend a lot of time visiting.

Only a couple of days after I move to Fort Erie, the Boyfriend picks a fight with me. He claims I am too clingy, that I do not give him enough space. He moves into a tack room in our barn and leaves me struggling to pay the hotel bill by myself. For a few days, we work in the same barn without acknowledging each other. I am not sure what has happened for him to turn on me this way. Slowly he thaws towards me, but he does not move back to my hotel.

The Boyfriend has acquired half ownership of a two year old filly. He surprises me by coming around on the horse and offering to let me gallop it. He loans me his helmet and leads me out to the training track where I can ride unobserved. I am so nervous that I gallop the first half-mile holding my breath. Then I relax, breathe and the horse goes fine for me. I am quite pleased with myself.

The Boyfriend is not pleased at all. He brusquely tells me I am no good. I had my reins too loose, I should have pushed the horse up into the bridle (whatever that means; he does not bother to explain). To prove his point, next day, the Boyfriend schools a young horse on the grass outside our barn. He rides the race horse like a dressage horse, collecting the animal up and riding around in circles. The talent of the Boyfriend is obvious to me and I hope one day to ride that well. I beg for another chance to gallop and he lets me ride the filly a couple more times.

Later that week, he comes by and asks me to pull a mane for one of his clients. I can use the extra cash. I find the mane has been neglected for a long time, having grown long and tangled. I work on the horse's mane for quite a while but the outcome is not as neat as I would like. When the Boyfriend shows up to check on me, he is furious. He launches into an angry tirade.

"I can't believe you have made such a mess of this. This is terrible, terrible. I said you would do a good job but now look at this. See how bad you have made me look. How can I put you on horses when you can't even look after them properly?" he screams at me.

I am dumbfounded at his behaviour. While the mane is not perfect, I still consider it better than it was before. It was so neglected that it was difficult to tidy up properly. In a month, it will need to be pulled again and will come out better. The Boyfriend will not hear anything in my defence. I feel that I have let him down somehow. I do not have the courage to ask to be paid for my labour. His erratic behaviour leaves me mystified, mortified and bewildered.

We alternate between this pattern of delicious courtship and touchy spats. One minute we are cozy lovers, next thing I know something has set him off and I am peacemaking. What is truly puzzling is that some of his friends obviously dislike me and I have no idea why. Sometimes, when we visit friends of his, he is warmly greeted but I am either ignored or coldly received. I once overheard one of his friends say, "Don't bring her back." For the life of me, I cannot figure out what I might have said or done to annoy them.

    The Boyfriend likes to put on a good show for his friends. He plans to take some of our closer friends and me out to dinner in Toronto. Suddenly piqued at me, he takes his friend's wife out instead of me. I am hurt but pretend this does not matter in front of everyone else. When he shows up for dinner without me, our friends ask where I am and he concocts a story. Next morning, I find him in his tack room, feeling rather miserable. He confesses he did not have a particularly good time. He decides to move back in with me.
    A few day later, the Boyfriend and I get into another terrific fight. He says he does not want me to watch him as he shaves. I think this is silly. I used to watch Dad shave when I was a child and found it quite comforting. Our fight quickly escalates until he starts to walk out on me. Crying, I grab his arm, trying to prevent his leaving.
    He begins to scream, "I can't breathe! I can't breathe!" His face has gone beet red. When I release his arm, the Boyfriend rushes to the window for some air. He sucks in great gulps of air. Then he turns around, grabs a bread knife sitting on the table and advances menacingly toward me. I easily knock it out of his hand. Abruptly, he disappears out the door. I am left sobbing on the bed, abandoned. This time I let him go.
    I am sick about the fight. While I acted childishly, his reactions were unbelievable. Yet the fault is all mine. Nothing I do is good enough. Then all is forgiven. Mentally, I am like a ping pong ball, bouncing around, trying to figure out what I did that was so wrong.
    I am always ready to forgive him, to blame myself, to try harder. I list my faults in my journal and plot how I can do things differently in the future. I am determined to become a more even tempered, likable person but I do not know how yet. I am sure I will figure this out.

I also know the Boyfriend has terrific potential and I want to support him to achieve it. It does not occur to me that he is perfectly aware of what he is doing to me, playing my desire to help against me. It is my nature to try harder but the game keeps changing.

Abusive behaviour is insidious. The last person to catch on to the game is the victim. What I do not understand is that the Boyfriend is a psychopath, who is fully aware of the control he has over me. I am the mouse to his cat, someone to be toyed with. His secret goal is to break me.

A few days later we start speaking to each other, cautiously feeling each other out. We resume our former relationship but it is just not the same. I have crossed an invisible line. Everything conspires against me. Now even my cat deserts me, going feral at the track. I am at low ebb.

On the last day in Fort Erie, after the meet had ended and most people have shipped back to Toronto, I give up my hotel room and sneak into his tack room for the night. With no one around us, we talk quietly.

"Why don't you sell your car? Then you can help me pay off my car," he suggests.

I think this is an odd request. I use my car all the time. Everything I own is packed into my car at the moment. What benefit is there in this plan to me? I do not even consider the request seriously but it stays in the back of my mind.

When Fort Erie's meet mercifully ends, I move back to Toronto and into the dormitory at Woodbine. I have not accomplished much this summer except sneaking in a few gallops. I walk hots for another trainer for a few weeks, which gives me more time off in the afternoons. For once, I am glad to have some distance away from the Boyfriend. Meanwhile, I keep my ear to the gossip network and put the word out that I am looking for a job that includes some riding.

Chapter Ten

UNDERVALUED

I finally learn of a farm north of the city looking for help, some riding involved. The job was given to another girl but she has nowhere to live so she turns down the job and offers it to me instead. I find a room to rent close to the track. I also sign up for psychology night classes. I hope that if I understand the source of my problems, I may not be so influenced by the Boyfriend and make progress towards my career goals.

After the fiasco of the summer, I am also determined to put more distance between us. He always claims I am smothering him. Maybe if I can stand on my own, not fall prey to my own emotions, we can work out our difficulties.

The farm that hires me is associated with one of the better known stables. There are two barns, with both an outdoor track and indoor track. The breeding barn has a few stallions plus the pregnant mares. The other barn houses the training division. The person in charge is also the trainer of the racing stable at the track and spends most of his time at Woodbine. A genial young Irish man named Sean oversees the training division.

There are four girls plus an elderly man named Bill already working in the training barn, now full of yearlings to be backed.

Except for Diane, they are extremely pleasant to me. Diane, however, seems to have a real chip on her shoulder and takes pleasure in being difficult with me.

Since I am the newest help, I have to muck the barn on Sundays with Bill, while the rest of the girls have the day off. I am allowed to choose a weekday for my day off and I decide on Wednesday. As the person with the least seniority, I only get to ride one yearling per day. I look forward to my single ride. The two senior girls begin riding while the rest of us are still mucking stalls. By mid morning, all four of them are riding. When my turn comes, Diane rudely dumps the saddle at my feet. I bite my tongue, praying in time she will warm up to me.

The first morning, as the four of us ride out to the track, the girls explain the riding procedure. We pair up with the two senior girls in front. At a signal from them, each of us from behind pulls out and passes, taking the lead. Once we are in front, the ones behind leapfrog with us. The girl beside me accelerates to the lead and I follow. Soon we are flying. No one is slowing down! I cannot believe we are at a full out gallop on yearlings. I have been part of a breaking crew twice before (during my Humber years I also broke yearlings) and we were never allowed to move beyond a gentle canter. On the way back to the barn, I question the other riders, who assure me this madcap riding is a daily event. Indeed, Sean seems pleased by our performance.

In the afternoon, all the girls start giving the yearlings a full groom. After an hour or so, I hear stall doors closing and the barn goes quiet. Curious, I pop into the tack room to find a game of cards in full swing, with Sean joining them. I am invited to play.

This does not sit well with me. Less than half of the youngsters have been groomed, yet the others are perfectly comfortable to spend the rest of the afternoon having fun. I thank the girls for their offer and go back to work. This card playing occurs daily. At the end of the afternoon, feeding commences. Each full tub contains ten different grains and additives. A five gallon pail of hot water is dumped into each feed bucket by one person and immediately carried into the stall by another worker. This seems a bit strange.

Why so many feedstuffs and why is so much water then poured over the grain? Since I have been at the track, I have learned not to openly question the odd practices that I see but to observe how effective different management methods are.

Sundays are very different. There is only old Bill and me to work and no riding is done. The frantic pace of the week slows down. After feeding and turning horses out, Bill and I have most of the day to muck stalls and no one to oversee us. After a while, Bill and I lean on our forks and Bill begins to talk with me. I quickly learn what a gem this aged horseman is. Diane does not get along with Bill. When she did the Sunday mucking before me, she snubbed Bill as not worthy of her notice, just an old man doing odd jobs. However, when he tells me who he galloped for as a young man, my mouth falls open in astonishment. Bill has spent fifty years in the thoroughbred industry, galloping for some of the legendary stables in the U. S. He has many injuries to show for his knowledge. Having worked so long in the business, Bill has developed a keen eye for the management of horses but the wisdom to keep his opinions to himself. When he finds I am a willing listener, he begins to point things out to me.

"Have you ever noticed, Jan, what terrible coats the yearlings have? The water is supposed to be absorbed by the grain for half an hour. The way the girls feed it, the water just washes the vitamins right through the horses."

That comment opens my eyes. All of the yearlings have dull, furry coats and lack muscle tone. They should be a picture of good health. Despite the incredible amount of good feed they were eating, the yearlings are not getting much nutrition from their food.

Another time Bill tells me to watch for unusual lamenesses due to our training practices. "Sean does not know anything about training. He is actually a broodmare man."

Bill is right. Three of the yearlings develop splints on the outside of their hind legs. Splints are a fairly common occurrence in horses, usually due to strain or sometimes a kick from the opposing leg. Splints, however, occur on the inside of their front legs. Eventually I am able to find a rare reference to splints on hind legs in an archaic veterinarian tome.

By coincidence, one of the yearlings that has come in for breaking is the youngster we had so much trouble catching when I worked at Aurora Meadows. The girls are having trouble backing the colt, now named Wayover, because he is so nervous. I watch them as they try to gentle this colt but they are not having much success. He blows up and bucks on the slightest provocation.

"Why don't you sack him out?' I ask.

Sacking out is a time honoured method of desensitizing a horse. Using a towel, you rhythmically tap the horse over his entire body. You start lightly in an area of the body where the animal is most used to touch, letting the animal flinch, expanding until you can reach almost every part of the body. Soon the horse stops reacting and you gradually step up the intensity of the sacking out. The horse learns he is not being hurt and the rhythmic motion relaxes him. It is especially useful on these nervous horses. Sean immediately defends his approach and continues to struggle to get the horse to accept the rider. I can only shrug my shoulders and walk away.

Once we finally get Wayover going under saddle, he continues to be erratic. Some days he will gallop nicely, other days he will explode and run off uncontrollably. He is such a good looking colt that I am sure his owners have high hopes for him.

Speaking of owners, I do not recall any of the owners coming out to visit their expensive investments. I consider the few weeks starting a horse under saddle as the most important of his life. If anything goes wrong now, it will be nearly impossible to correct later. Yet here we are, madly galloping about on these immature babies with no protest from anyone. Even the barn's trainer never shows up to watch.

I try to fit in with the crew but Diane continues to treat me as rudely as she dares. Then, one morning, Diane arrives a couple of hours late. We learn she has been in a car accident and was caught driving with her license suspended. I feel a certain smug satisfaction that Miss High-and-Mighty has been taken down a peg or two.

She pointedly avoids me but I prefer this to being picked on constantly. As I begin to understand human nature, I see she treats me the way she does because she is so insecure in her own status and skills. I vow never to treat others the way she has treated me.

The farm owns some of the yearlings and is sending them to the track for a couple of weeks for some exposure there. Just six weeks after I left the track, I return as a groom for the yearlings. Now that I am back on the track, I start spending more time with the Boyfriend.

I have been playing it cool with the Boyfriend, refusing to be hooked in to his manipulations. Well, some of the time. After a couple of weeks of keeping my distance, the Boyfriend starts seeking me out. Yet at other times, he lies to me. One day he and Donny are leaving the track and I ask to go with them. The Boyfriend claims they are just going to the store but they drive to the equine hospital at Guelph University instead. I have never been and would have loved to go. Donny refuses to answer my question about why he went along with the lie, preferring to stay out of our arguments.

Now, the Boyfriend decides to go to Florida to work for the winter. My hope is that the physical distance between us will bring us closer. A distance between us has always worked magic before. He asks me to come down for Christmas but there is no way I can afford to. We have an emotional parting. He drives down, taking his friend Donny with him.

After a couple of weeks at the track, the yearlings are shipped back to the farm. Now it is November, the leaves have blown from the trees and my prospects are as bare here as the trees. My gut feeling is I am about to be laid off. With the breaking winding down, I am sent over to the breeding barn to work. Even if I do work throughout the winter, there does not seem to be any chance of doing more riding. Also, I am beginning to dislike working for bigger outfits where my opinions do not count for anything. I decide to keep my eyes and ears open for a better opportunity.

Chapter Eleven

Scottish Music

My opportunity to ride comes with a new trainer at Woodbine. Russell has just been at Woodbine a few months. When he is about to move his horses home for the winter, his exercise rider quits rather than work on a farm. I locate Russell at his stalls at Woodbine. We get on so well that he hires me on the spot. He promises to help me with my riding and I take the job without hesitation.

The farm I am going to work at is located about half an hour west of the track and belongs to a wealthy businessman. The owner, J. B., has built himself a new house on the farm. Russell and his family have moved into the old farmhouse.

I will no longer have to pay rent since there is also comfortable living quarters attached to the barn for me. It is a large room with a full kitchenette on one side with a private bathroom. On the other side is the laundry facilities shared with the barn. Through the laundry room, I can walk directly into the barn. The living accommodation is ideal in every way except one; I am pretty lonely. The Boyfriend has gone south and I am forty minutes away from Toronto, where my friends live.

Russell's wife guesses at my loneliness and invites me up to the house for several meals and watch television afterwards. After a week of eating meals up at the house, she and I work out an agreement for me to pay board. In return, I am welcome at the house at any time and soon become friends with both Russell and his wife. This assuages my loneliness.

Russell plans to keep several horses in training over the winter and be ready to race as soon as Greenwood opens. In the first **set**, I ride Snuffy, an easy going three year old gelding. Surprisingly, J. B., who is long past his sixtieth birthday rides with me. He is mounted on Spike, a two year old colt. After that, we ride a pair of fillies. Then J. B. bids us good day and leaves. Russell and I walk up to the house for coffee break.

Before we leave the barn, Hank, my next mount, lies down for his nap. This three year old is so relaxed, I can walk into his stall and sit on him without the horse even standing up. No doubt he is saving his energy for his workout.

After coffee break, I saddle up Hank. Russell can be relied on to put the Scottish music on the P.A. system. And somehow, there is always an audience there to watch me. Friends of Russell's, his wife, whomever.

Hank is truly one of those characters with a sense of humour. His favourite gambit is to listen to the drum roll on the Scottish music tape, which is guaranteed to start Hank **crow hopping**. It does not take much to send him into a spree of bucking. He never bucks hard enough to drop me but just enough to keep me on my toes. My audience loves watching me ride Hank's challenges.

As the weather gets colder, I bundle up to stay warm as I ride. I have the hood of my sweat shirt pulled up over my head with my helmet clamped down over it. While this keeps me fairly warm, it does limit my peripheral vision. One morning, as I am concentrating on staying abreast of Spike, a horse hoof suddenly lands in my lap. As I glance sideways, I am horrified to see J.B. rolling over and over in the dirt. Spike has tried to mount Snuffy while we are cantering together. J.B. is too old to be falling off any horses. Although he professes to be unhurt, Russell switches us around and I ride Spike after that.

During this time the Boyfriend phones me regularly from Florida. Our connection is getting stronger. I always know when the phone in the barn rings if it is him. Sometimes I am holding out my hand for the phone before Russell realizes who is calling. The Boyfriend says he misses me and that Florida does not suit him. He tries to talk me into leaving this job and joining him in Florida. I have no intention of leaving this job; I am gaining confidence in my skills and am guaranteed galloping at Woodbine next spring. I also like being the only employee and all the responsibility that comes from this.

One morning we **worm** all the horses on the place. This is before paste wormers become widely used. Instead, we have the vet coming in to tube worm the horses. Literally, a rubber tube is threaded into a nostril, passed down the esophagus and the liquid poured straight into the stomach. Russell has 'drawn' the horses the night before, meaning, he had taken away their hay to allow the wormer to be more effective. After worming, we hay all the horses.

Within hours, it is evident that Spike must have stuffed himself with his straw bedding last night and is colicky. He is restless, pawing and refusing his hay. While he is not very bad, we still have the vet back to treat him. Colic is a major cause of death in horses and should never be taken for granted. We keep an eye on Spike for the rest of the day.

At five o'clock, just as I am getting ready to go out, Russell knocks on my door.

"Spike is colicking again. I need you to help."

I sigh. I will not be going out tonight.

The two of us are up all night with the horse. Spike is not in extreme pain but obviously, his gut is blocked with his bedding. We take turns walking him. The vet is called back to treat the ailing animal. The horse becomes weaker and it looks like we might lose him. At five in the morning, Russell gives Spike a mineral oil enema. Finally, the blockage begins to soften and is passed. The horse will live. But it was a close call. I go to bed for a few hours and catch some sleep.

One other incident stands out from that winter. One of the yearling colts goes off his feed. When Russell checks the colt, he finds a front tooth split in half. I am the one who must hold the little guy for the vet. I can handle gaping, bloody wounds without flinching, putting down a horse when the occasion calls for; assist vets in all sorts of emergencies. But teeth issues I am squeamish about. I have always had horrible teeth and not the most tender of dentists. It is not a fun afternoon for me.

Our best guess of how the colt has hurt himself is that the bars in his stall do not reach all the way up to the ceiling. This horse sometimes amuses himself by standing on his hind legs and running his teeth along the top of the bars. He must have slipped, splitting the tooth.

I find Russell very congenial to work for. Once that winter, Russell oversteps his bounds and makes a mild pass at me. I let him know I am not the least bit interested in having an affair. Besides, I like his wife too much. Russell takes it in good order and the event is put behind us.

Hank has finally started to settle down and train well. Because he has already raced, he knows how to gallop properly. Since I ride alone when riding him, I can practise my newly learned galloping skills. With my stirrups shortened, I stand up and balance myself by resting my hands on Hank's neck. I have the reins doubled in a **bridge,** gripping them and pressing my knuckles into Hank's coat. I am becoming adept at keeping my balance and influencing Hank's pace. For weeks now, Hank has been leaping about and I have stayed glued on top of him. I am feeling pretty cocky about my skills. Now, when I least expect it, Hank simply makes an unexpected deke. I never see it coming. One moment, I am privately congratulating myself on how well I am doing, the next moment the horse isn't underneath me. He simply makes a left turn.

When I face plant into the dirt, I think Russell is going to fall down, he is laughing so hard. Horses sure have a way of bringing people down to earth.

Scottish Music

I am feeling pretty settled in this job and plan to stay several years with this outfit. When I refuse to join the Boyfriend for Christmas in Florida, he returns instead, claiming to have missed me. He has only been gone six weeks. Since the Boyfriend does not have a job, Russel hires him to ride the two fillies that have just gone into work. I am relieved that I still have all my regular horses to ride and pleased that Russell is so loyal to me. I also feel more equal to the Boyfriend instead of putting him on a pedestal.

The Boyfriend moves in with me and for a week, everything is cozy between us. I quit eating at the house and start cooking for the two of us. Then the Boyfriend picks a fight over what I serve him for supper and walks out on me. I am completely stunned. I can't believe he is leaving me so soon after coming home. I suspect that the real reason he is leaving is because Russell has not given him all my mounts to ride. Why the Boyfriend would be so jealous is beyond my comprehension but this time I do not cave in and chase him. I am standing firm against his behaviour. In a few days, I hear that he has moved back onto Woodbine and is free lancing again.

I drive to Woodbine and find the Boyfriend. We have a long talk. He is apologetic for the way he behaved. Since he is so contrite, I confess that Russell made a pass at me. We decide to put all that behind us. Now that I am not so reactive to the Boyfriend's bad moods, he seems intent on having a future together. He even mentions marriage. I am pleased. Our future is looking rosy again. What I do not know is that he is already seeing another woman.

Russell and I make plans to move into Woodbine in mid February. Then, a few days later, when I am relaxing in my room, Russell knocks on my door and comes to talk with me. He exchanges some pleasantries, unsure of where to begin.

"I have bad news for you. J.B. was talking to me this morning. He said, "Get rid of Jan."

I tried to talk him out of it but he would not budge."

I am completely taken aback. While not chummy with J.B., I can think of nothing I have done that would even come close to getting me fired.

We talk for a few minutes more. I am trying not to cry, I am so disappointed and hurt. Russell is genuinely sorry but his hands are tied.

As he gets up to leave, he glances out the window. "Oh, look, a deer." He points at the animal. "That's a good omen for you."

For some reason, the deer makes me feel a little better and my optimism returns. I will get a galloping job, of that I am determined. Neither J.B. nor the Boyfriend are going to stop me.

What I learn later is the Boyfriend is behind me getting fired. He actually brags about getting me fired. When I had told him about Russell making a pass at me, he tattled to J.B. to spite me. Behind my back, the Boyfriend is doing everything possible to keep me dependent on him, or at least unsuccessful on my own.

I am only just beginning to see the pattern here. To be successful, I am going to have to cut my ties with the Boyfriend.

Chapter Twelve

NO SENSE OF HUMOUR

In the good old days (well before my time and when girls had 'ruined' the track), the racetrack was a much easier going place. There were poker games that took place in the kitchen which had trainers winning and losing thousands of dollars in a single game. The grooms really knew their jobs back then. Why, a person had to hot walk two or three years before he was allowed to so much as touch a horse. The old timers knew how to keep a horse sound with their secret remedies and used specially cooked feeds to keep their charges at the peak of his condition.

This was, of course, before grooms were allowed a day off in the week or made enough money to afford a car. Dental plans and Worker's Compensation lay far in the future. The grooms lived in the stalls beside the horses in their care. There were no cooking facilities for the people, only for the horses. Of course, there was only eighty seven days of racing, compared to the two hundred plus days of racing in our meets.

Once upon those times, the Jockey Club held hotwalker races. These were no measly affairs. These were sprints open only to bona fide hotwalkers and were held in front of the grandstand between the regular races. The prize money to the winner was one hundred dollars, more than a week's wages.

The backstretch folks fully embraced the fun. The grooms acted as trainers, carefully conditioning their hotwalkers. Great secrecy surrounded the training. Alliances were made, plans finalized. No stake horse ever received better preparation.

When the Great Day arrived, the hotwalkers were swathed in **rundown** bandages and wrapped in the stable's **coolers.** The hotwalkers were paraded in front of the grandstand, then led down to the gate.

Where the money goes, so does corruption.

One year, Jake the Rake planned to win. No one noticed as his groom hid him in the bushes at the top of the turn. No one missed him in the starting lineup. He let the others race past him, then jumped out from the bushes and joined the runners. He made a come from behind and won easily. He accepted the prize money and no one, for the moment, was the wiser.

A windfall of money to a poor man is like a feast to a starving one. He blew all the money at the bar that night. Unfortunately, when he got drunk, he bragged about how the race had been fixed for him to win it.

That was the end of the hot walking races.

Ah, the Jockey Club has no sense of humour.

What is the Jockey Club? The Ontario Jockey Club is a private organization which governs racing in the province. It has done so since legendary business man E.P. Taylor reorganized racing here in the 1950's. The original jockey club, started in Britain about 1750, was a gentlemen's club formed to establish the rules of Thoroughbred racing. Our Jockey Club is affiliated with the American Jockey Club, who registers some 25,000 thoroughbreds each year. By unifying the Canadian branches with the American, the same rules and licensing requirement are used and it is easy for horses to cross the border to race. The racetrack is ruled by the stewards, the officials who are judge and jury. If you are called upon the carpet, you know you are in deep doo-doo.

The life line of the backstretch is the P.A. system. It calls us over for races and afterwards to the test barn. It announces the results of the races. Few trainers have phones installed in their barn, therefore they are paged to the office or hear which of their help is sick and not coming in today. From the P.A. system, we learn if a jockey is overweight and taken off their mounts for the day, where the vets are needed and how the entries are proceeding.

There is an auxiliary mic at the loading ramp which calls us there when we are shipping out for the races. One morning, a groom arrived at the loading ramp with his horse only to find the stall man had not yet arrived but the mic is turned on. This is way too good an opportunity to miss.

"Okay, all you f***er's" he began. He continued in this vein for several minutes.

The backstretch was in stitches. The reaction in the office was closer to apoplexy. Subsequently, the mic is locked up when not in use.

The Jockey Club didn't think much of this either.

Then there was the incident of The Weasel. The Weasel was a successful but mean spirited trainer who would claim a guy's last horse. When he was short of help, an owner-trainer with only one horse stabled down from The Weasel helped him out by walking some hots. The Weasel repaid him by claiming the trainer's only horse. It is understood around the track that you never claim a horse in your own barn or take a guy's last horse.

Of course, there was tit-for-tat. When the previous trainer returned to the paddock with the horse after the paddock, he found only The Weasel waiting, instead of a groom. The Weasel was without a halter and shank. The trainer was so mad that he simply yanked the bridle off the horse, leaving The Weasel trying to keep the horse contained in the stall until his groom arrived. The Weasel was forced to leap back and forth, waving his arms to keep the horse contained. He would have looked pretty silly. Too bad I missed that one.

Not many girls worked for The Weasel. He did have an exercise girl working for him until he made a pass at her. She quit on the spot and went down to the kitchen to assuage her feelings over coffee.

Not long after, The Weasel headed down to the kitchen for breakfast. As he sat down, the exercise girl spotted him. She returned to the cafeteria lineup and selected a nice cherry pie slice. She calmly walked over to The Weasel and pasted him with the pie in the face.

It caused quite a stir in the kitchen. The Greek who had sold her the pie rushed over and said, "No charge. No charge."

Yes, the Weasel was not very popular.

At Greenwood, The Weasel had an annoying habit of parking his car directly between the barn door and the manure pile. The grooms were forced to carry the manure around his car each time.

Russell was stabled next to The Weasel at that time. One morning, The Weasel drove in with a brand new car. He parked in his usual spot, then buggered off to the office.

It just so happens that Russell had **physicked** three of his horses the previous day. Physicking is an old time remedy of dosing a horse with a medication that cleans out the intestines. It produces copious, vile smelling diarrhea the following day.

Finding the car in his way, Russell opened the nearest car door and proceeded to stuff the car with the slimy, smelling manure. He continued until the car was full.

The Weasel was furious. He immediately went to the Stewards about Russell's prank.

And the Stewards just laughed.

Maybe, just maybe, the Jockey Club has a sense of humour after all.

Chapter Thirteen

AT THE GATE

I have come back to the track in mid February to apply for a job as an exercise rider. I want to gallop so badly I can taste it. I'm no longer willing to settle for anything less.

Being fired unexpectedly, I am now scrambling for one of the few openings for an exercise rider, with no experience behind me. If I could have stayed with Russell, I would have had a built in position galloping the very horses I rode on the farm. Instead, here I am once more in Jimmy's trailer begging for an opportunity.

Jimmy hems and haws. After a few minutes deliberation, he makes some phone calls.

I wait uncomfortably in the barren trailer for half an hour or so, until Jimmy says he might have something for me. We climb into his golf cart and I am whisked once again into the back stretch.

Jimmy hails Doc and introduces us. Doc appears elderly, with a bird-like quality about him, like a vulture with his head jutting from his skinny neck and his claw-like hands. I am mildly repulsed by his appearance but I want the job.

Doc asks some questions about my experience, then asks me to try out a two year old. With Doc on his tall gray thoroughbred Maxwell and the other rider Cathy on another filly, we head over to the sand ring to gallop. Lurullah is an almost pony sized chestnut filly, very sweet and very easy. When we return to the barn, Doc has me jump on Licorice, a leggy two year old colt. Licorice is a little more of a handful but also goes nicely for me. After I gallop him, Doc officially hires me.

Doc is one of the few trainers whose help both rubs and rides. We muck the barn, take a coffee break and start riding. We cool each set out after we ride, tack up the next set and continue until done. We rub off our horses and clean our tack. I am in heaven.

Next time we see Jimmy driving by in his golf cart, Doc stops Jimmy to thank him for sending him such a good worker.

When I drop by my previous boss Gordon's barn a couple of days later to say 'Hi', he asks who I am working for now. When I tell him, Gordon just replies, "Oh-h." No explanation.

Out of necessity, I move back into the girls' dormitory. The following week, the manager of the dorm has a coffee klatch in her room so we can meet the other girls living there. I become friends with Sandy, an easy going strawberry blonde.

I quickly settle into my new job. My co-worker Cathy comes from a background of show horses. She rides beautifully with an elegant seat and quiet hands. I am more a seat-of-the-pants kind of rider, gutsy, determined and often just as effective. Being newbies in training, we work and learn, side by side. Through encouraging each other through our difficulties, we become fast friends.

Cathy is such a cute, vivacious blond that all the guys on the shedrow fall madly in love with her. hotwalkers and grooms go out of their way to come by our stalls each day to say 'Hi' to her. With her modest personality, she is equally well liked by the women.

For the first couple of weeks that I gallop, my back is in agony from muscle spasms. I have had my share of falls from horses and they have taken a toll on my body. I am not sure if I can manage this. I tough this out and grit my teeth to continue. Then my back gets stronger and the pain goes away.

Doc continues to ride out each set on Maxwell, with us on either side of him. Doc sets the pace and looks out for us as we learn to handle our mounts. "Horse coming up on your inside," he warns, or "Watch for the two horses working out of the gate."

The horses themselves teach us the most about galloping. On just my third morning, Licorice digs his heels in at the five eights pole, makes an abrupt left turn and high tails home without me. A week later, he repeats his performance – with the same result. I walk home again, enduring the embarrassing jibes of the other riders.

The Boyfriend seems surprised that I have landed a galloping position. Now that I am back at the track, we spend most of our time together when not working. The Boyfriend still manages to have people around us. He introduces me to Jaylee, an American girl rubbing some horses where he is freelance galloping. Because she is new here and does not know anyone, we take her with us to dinner a few times.

One morning, I stop by the Boyfriend's tack room before work. I walk in to find Jaylee just getting dressed. I am so shocked I do not know what to say.

I decide to break off with the Boyfriend. While he is working, I remove my few possessions from his room. I try to avoid him after this.

A few days later, the Boyfriend calls me to his tack room. We have a talk and he promises me it will not happen again.

A few days later I see Jaylee driving his car past my barn. There is no reason to be driving near my barn. I suspect she wants me to see her.

I decide to confront her and tell her where things stand between the Boyfriend and me. I call on her at her room in the dormitory and lay the facts before her. The next day, when I see him, the Boyfriend starts screaming at me that he no longer wants anything to do with me.

Occasionally, since we both gallop, my path does cross that of the Boyfriend. He never hesitates to embarrass me by yelling some less than complimentary remarks about my riding. I try to stay out of his path.

Donny returns from Florida but he is not speaking with me. I have no idea why.

In just a few weeks, the bitter cold is replaced by milder weather. As the frost comes out of the track, the spring rains begins and the track becomes a soggy mess.

Galloping in the mud is a new experience for me, as well as for the two year olds. I am riding Licorice, totally unsuspecting, in the first set. We pick up a trot and an eighth of a mile further, break into a canter. As I stand up into my gallop position, balanced on my stirrups, Licorice spread eagles underneath me. I somersault down his neck, pulling the bridle off as I fall. Doc and Cathy look back for me but there is nothing they can do. As they continue to gallop, I go looking for my errant mount, who has exited via the **in gap**. A groom in the adjacent barn catches him. When I get a leg up, I ride back out to the track.

There is an old race track saying that a certain horse 'can't stand up in the mud' meaning he runs poorly on an **off track** but Licorice literally cannot. He is slipping and sliding like Bambi on ice. The only way I can get this gelding to balance himself is to carry my hands up by my chin and jerk on the reins whenever he tries to balance on my hands. It is nightmarish but we get around somehow.

Two days later, he falls again with me. I tumble onto my right side. Later that morning, I am cantering Lurullah through the same spot. Because she is bucking shins, she doesn't switch her lead and she goes down. I land on my left side. When she doesn't move, I believe for an agonizing few moments that her leg is broken. She must have just been winded because she soon recovers.

By the time I ride back into the barn, I am covered with mud from head to toe. One of the groom razzes me, "You didn't fall off again?"

"I didn't fall off," I protest. "My horses fell down."

"Sure, sure."

After this, broke as I am, I scrape the money together to buy a rain coat and pants for galloping in muddy weather.

Doc surprises us one day by telling us we are going to give Licorice and Brownie T (Cathy's mount) their first quarter mile workout. Cathy and I are so excited. While Cathy has worked a horse before, I never have.

Doc rides out with us to the in gate. He calls up to the **clockers**, "My Licorice and Brownie T to work one quarter." He parks Max on the outside rail by the finish wire.

Cathy and I gallop once around to warm them up. On our second trip, as we approach the quarter pole, we move our horses down to the inside rail. Suddenly we smack our surprised mounts with our sticks and Cathy starts to hiss like a snake. I, too, am yelling to urge the youngsters on. I squat low, my belly touching my thighs and my arms are thrusting forward with each stride. At first, the response is sluggish but then they begin to actually gallop. The action of the canter changes from a fluid rolling motion into a piston-like pounding. I am aware of nothing except the wind in my face, the pounding gait and Cathy beside me.

At the finish line, we stand up and let them gallop out another eighth of a mile. My legs feel like Jello and I think I may throw up with the exertion. No wonder jockeys look so exhausted after a race!

The Boyfriend was watching me and yells at us for riding with loose reins when working our youngsters. While grateful for the instructions, I hate the way he publicly humiliates me.

Disappointingly, our time was too slow to make the **time sheet** at twenty seven seconds. Fortunately, first workouts are not much of an indicator of the talent of your horse.

After that, the jockeys work our horses over the longer distances. I am secretly glad. I can do without feeling like Gumby.

After he has worked a couple of times, Licorice begins to get strong as he approaches the quarter pole. I struggle with him until we pass the finish line, where he eases up of his own accord. One morning, as I am approaching the quarter pole, Doc says quietly, "You'd better move out. There are four horses working together."

Under no circumstances are four horses supposed to work abreast. They pose an unnecessary risk to other horses. Morning training is not racing.

As I glance over my shoulder I realize I am going to be in trouble. Sure enough, as the four horses fly past us, Licorice bolts after them. I stand up in my stirrups for more leverage and seesaw on the reins but it is no use. I give up and fold back down on his neck, tucking into the rail behind the others.

Once on the backstretch, Licorice begins to pull himself up. "Oh, no, you don't," I hiss angrily. "You wanted to go, then I'll bloody well make you go." I kick him in the ribs and force him on. A mile further on, the colt is well and truly ready to stop.

I join Doc and Cathy, waiting for me. "That was about 23.3," Doc comments with a slight smile. Licorice has carved about four seconds off his time.

The next day, Licorice works the half in 48.2. He is beginning to show his talent.

After a day of rest, it is time to gallop again. Doc digs up an overcheck bridle for Licorice to teach him some manners. This piece of equipment is usually reserved for **Standardbreds.** I have never seen it for a Thoroughbred.

When we hit the quarter pole and Licorice tries to run, he grabs the bit for three whole strides before giving up the idea of running off. The more severe bit works and now I have control. He gallops like a gentleman the rest of the way.

Racehorses must do most of their work at a slower pace, otherwise they will break down prematurely. Even within a race, the horse must depend on the jockey to set the pace and decide when to challenge for the lead.

Doc often kindly has our coffee waiting when we take our mid morning break. No matter how much I protest, Doc always forgets and heavily sugars it. Normally, I drink my coffee without sweetener. Later, I learn the sugar is probably heavily laced with hormones. Doc must be hoping to 'get lucky' with one of us. Is it any wonder I am struggling to break up with the Boyfriend?

Deciding to leave the Boyfriend and actually finding the willpower to do so are very different things. My heart still leaps when I see him. My body aches for him. In this past year, I have forged a psychic connection with him, a sense of knowing when he wants me.

The more I walk away from the Boyfriend, the more he chases after me. Sometimes I wish he would just leave me alone. He will send for me, we'll spend some quality time together and than I will catch him in a lie. This push-pull is taking a toll on me.

Sometimes I take my new friend Sandy to the coffee shop and talk her ear off with my problems. Cathy, too, is supportive at work, although she leaves at noon with Doc to work at his farm.

Doc is pleased with the progress Cathy and I are making and decides he no longer needs to ride out with us. Cathy and I are excited to be allowed to ride without him. We set off galloping carefully, keeping exactly to the same pace as Doc always set, watching out for other horses that could cause us problems. We are very pleased with ourselves as we ride back into the barn. Doc is not. He calls us into the office to have a little chat. Imagine our surprise when he reams us out for galloping too slow! What on earth were we doing?

Cathy and I look at each other and shrug our shoulders. There is no point in arguing with him. We open up our pace to stay out of trouble.

Doc's 'pony' provides comic relief to our endeavours. Maxwell is a grey field hunter, 16.2 **hands** high, good looking and a blithering idiot. He is spooky and completely unreliable, often worse behaved than the two year olds he is supposed to sedately accompany. When Doc spanks Max for his unruliness, Max is sure to spook twice as bad the next time. Max is not above heaving himself backwards or wheeling round at something he had not glanced at yesterday.

There are a lot of trainers wanting more stalls for their racehorses. The **Jockey Club** will not give Doc a stall for Max. Rather than take Max back to the farm, Doc builds a partition in the feed room to divide it in half. Max is stabled just inches from the grain. Inevitably, Max breaks loose one night, tears down the partition and helps himself to forty pounds of oats! It should have killed him! Instead, his hinds legs are swollen in the morning and Max is high as a kite. He bucks Doc off that morning and saunters home just ahead of the boss.

Doc sends me back out on Max to gallop another two miles to take an edge off him. My arms soon ache from holding him back. It is all I can do to prevent the gelding from running off with me.

Doc likes to ride out with the jockeys when they are working our youngsters. While the jockeys are warming up down the backstretch, Doc positions Maxwell's bum against the outside rail right at the finish line. He is in an excellent spot to observe the workout, standing in front of the **clockers'** stand. This day, as Doc relaxes while waiting for his horses to come by, he drapes the reins on Maxwell's neck and folds his arms complacently across his chest. Since Cathy and I are not riding, we have come to watch the workout. We are quietly standing in the booth behind the clockers.

Unexpectantly, one of the clockers slides the window open. Maxwell surprises Doc by flinging himself across the track in a single bound, nearly unseating our trainer. Cathy and I cannot contain our snickering and just hope Doc does not hear us. Good old Maxwell has done it again.

One fine day in April, when we finish mucking stalls, Doc calls us into the office again. Exchanging glances, wondering what we have done wrong this time, we present ourselves to him. Doc announces, "Girls, I am sending you to the gate this morning."

Cathy and I look at each other, our eyes wide with surprise. I have never contemplated being allowed to break a horse from the starting gate. Doc gives us a lecture on what to do and expect. We tack up and head out.

Doc mounts Max and takes hold of Brownie's bridle, keeping Cathy close beside him. As per instructions, I have Licorice's nose on Maxwell's bum as we jog the wrong way of the track toward the gate. As we come up the **chute,** Maxwell spies a loose plastic bag blowing in the wind. It is an opportunity too good to miss.

Without warning, Maxwell flings himself backwards. Brownie wrenches free and Doc nearly falls off his horse. Maxwell refuses to lead past the plastic monster so Licorice kindly passes and leads the way to the starting gate.

At the Gate

"First time for the horses, first time for us," we tell the gate crew. Doc positions himself behind the gate to watch us.

A beefy young man walks toward me, taking a narrow strip of leather from around his neck. He doubles it through the bit ring on my left side. Speaking soothingly, he backs up into a narrow stall of the gate. Licorice hesitates, then follows carefully. Even with my shortened stirrups, I have to lift my feet to clear the ledge on either side that wedges us in. The man carefully closes the two halves of the gate in front of me while the gate behind us is left open. This is done for the first gate experience to prevent the horse from panicking in the enclosed space.

Horses are creatures of flight. Their one defence is to run away from danger. That these animals should so trustingly enter this claustrophobic space is astonishing to me.

The stall is barely wider than my horse. The sides come up even with my thighs and are heavily padded. Above are the bars, with the space open at the back. This allows the gate crew to nimbly climb over the horses in the gate to assist riders in trouble. The doors form a V, pointing outward with just enough space for Licorice's head. Through the barred doors in front of me, I can watch horses gallop across the chute.

As Licorice turns his head to watch the horses galloping, he pins my leg against the ledge. Out of nowhere, a gate crew member climbs to my side, reassuring me and showing how to jiggle the rein to keep my horse's head straight. I feel my horse relax beneath me.

Next to me, Brownie is loaded on my right.

We wait.

My tension is tightening like a screw. Then a voice from behind asks, "Ready?'

I loop the reins and grab a double handful of mane. "Yes," Cathy and I answer hesitantly.

There is a distinct click and the gates fly open.

Licorice glances up as if to say, "Interesting. These things open." I hear Cathy yell and I nudge Licorice in the sides with my heels. My horse takes a couple of steps and clears the metal monster. Cathy is just ahead of me. I urge Licorice on and we begin to canter after her. We begin to pick up speed and race alongside her.

It is then that I hear Cathy laughing. Like the air expelling suddenly from a balloon, we are releasing our tension. All the way down the stretch, as our two horses stretch out, we laugh aloud, high on life. This day is a highlight of my life.

Chapter Fourteen

Mistakes

My backstretch license now states that I am an exercise rider; I am feeling pretty high on myself. If there is one thing I am learning about horses, it is their ability to humble you.

For a trainer with a poor record of producing winning racehorses, Doc has an amazing ability to find new clients. We receive three nice, well bred fillies from Quebec.

The dark filly, Coon, has extra white in her eye, giving her an unruly look. She also has three white rings around the top of her tail, probably from rubbing her tail at some point. This distinctive marking means she is easy to pick out on the track. Her **half brother** is running in the **Plate** this year. The owners hope she will be as good a runner as her brother. Doc assigns her to me.

The second filly, Chance, descends from one of the great Canadian race mares of all time. They also have high hopes for her. So Irish, the third filly, is the best looking but tends towards chubbiness.

On one of my first rides around the training track riding Irish, there is a small crowd of owners standing under the clocker's stand with their trainers. As I gallop on, I feel my filly hesitate. With more experience, I would have known to sit down and push her forward.

The way you become experienced is by making mistakes.

Irish suddenly plants her front feet and wheels away from the knot of people. I am unceremoniously dumped in front of them. There is nothing like having an appreciative audience to watch you make a fool out of yourself. The trainer kindly comes over and boosts me back in the saddle.

Coon turns out to be a pussycat to gallop, despite her looks. Chance, on the other hand, gives Cathy all sorts of grief.

This starts with the very first gallop. We have barely begun galloping when I hear a yell from Cathy and see her cartwheel through the air. Instead of running off, the filly stands quietly beside her. Cathy is able to get a leg up while I pull up and wait for her. The filly gallops a mile, then throws her off in exactly the same place. Again, she stands and waits for Cathy.

It becomes a morning ritual. Nearly every day, Chance will throw Cathy. These are not little stops in which the rider slides over the shoulder of the horse. No, I see Cathy spectacularly somersaulted over the filly's head. There never seems any obvious reason to set this filly off. Yet, the falls continue.

Cathy becomes afraid of the filly. When we ride out together, she is crying. Eventually, Doc is forced to put a jockey on the horse instead of Cathy.

Janice shedrowing a colt.

This filly has one other quirk. She is absolutely terrified of blacksmiths. We guess one of them has roughed her up once and she has never gotten past it. When the blacksmith parks his vehicle on the road in front of our barn, Chance is already pacing her stall, snorting in fear. It must be hellish for this filly to be shod. It is no picnic for the blacksmith trying to reset her either. Racehorses are shod religiously every four weeks.

MISTAKES

In our barn there is a three year old filly who Doc only puts more experienced freelance riders on. In the first few weeks when we are still using trouble lights to muck and groom with, she will not enter her stall unless the light is turned off. This filly, Traya, freaks out with a shadow passing over her eye.

Traya becomes one of my charges to groom. At first, her manners are pretty deficient. She doesn't hold her feet kindly to be picked or move over when asked. She isn't mean, just doesn't seem to understand. With lots of patience, she slowly improves. I begin to see her as brain damaged. She just takes three times longer than the average horse to be able to learn anything.

A groom in our barn tells me her story. As a two year old, when they first started to bath the horses in the warm weather, Traya was dancing around. Doc impatiently grabbed the hose and trained the nozzle directly on her. Her reaction to the strange event was to rear up and flip over backwards. The sound of her cracking her head on the cement reverberated through the barn. For three days she walked with her ears dangling sideways from her head. She is lucky to be alive. The incident must have caused her permanent brain damage.

Slowly, the trust grows between the little chestnut filly and me. I can get Traya to follow me kindly onto the transfer van when we run her at Greenwood. When the van passes under a bridge, the shadow it cast causes her to throw herself backwards. By talking soothingly to her, she calms down again.

One day, when she is entered in the afternoon's races, I am still busy in the barn when they call us to the transfer van. We send our new hotwalker with Traya. As soon as I have done up my horses and the others are fed, I drive down to Greenwood. However, when I walk into the receiving barn, I am told Traya is scratched from her race. She flipped over in the van. She missed my familiar voice to steady her when going under the bridges.

I am beginning to see that being a trainer does not mean you know how to handle horses, only that you can read the condition book and b.s. owners. The real strength of a barn lies in the unsung abilities of the help – the competence of the exercise riders, the dedication of the grooms, even the relaxed attitudes of the hot-walkers.

Another time, I am putting a bridle on Traya. We are facing the stall entrance with the screen closed. As I ease the bit into her mouth, all hell breaks loose. Traya throws herself madly and her feet go flying. I remember feeling her hooves on my back. The two of us crash through the metal screen, bending it badly. The incident happens so fast I am not sure what just happened.

The filly doesn't go far before she is caught. The bridle is dragging on the ground between her front legs, the reins still around her neck. We give Traya the morning off and have her mouth checked by the vet. The vet finds a tooth that gapes out from the rest of her teeth. Probably I have inadvertently banged the bit against it, causing a shot of pain that caused the filly to react.

One morning, Doc calls Cathy and I to his office again. To our astonishment, he gives us a raise. He is very pleased by the work we are doing.

This gesture of respect is in sharp contrast to my private life. My inner pain is wearing on me. There are days that I manage to stick to my resolution to cut ties with the Boyfriend and stay away. Then he will send word that he wants to see me. It is as if I am tied by an invisible string. Anytime crooks a finger, I come running. I know I will be better off without him but I just cannot seem to help myself.

Then things fall apart again and I know that I am wrong once more.

Through the grapevine, I have heard that Donny blames me for the Boyfriend dumping him in Florida. The two men have been friends for years but have fallen out in Florida.

Doc takes Saturday mornings off. Each of us in turn comes in on Saturdays, feeds the horses and walks each of them a few turns to stretch their legs. The other employees get the day off. Thus, I get two days off every three weeks.

One Saturday, when I show up at seven, an hour later than usual, I am puzzled by the sound I hear. It sounds like an excavation going on. I walk down the shedrow, trying to locate the source of the sound.

Mistakes

In Licorice's stall, I can just make out his bum, high in the air. Great hunks of dirt are flying up from where I imagine his head to be. Investigating further, I find his front legs are four feet below where his back feet are braced on the hill he has created. With no hay to keep him occupied, he is digging his way through to China. After feeding and walking him, I have to level out his stall as best as possible. I learn not to come in so late, even though Doc is fine with showing up at this hour.

Each Saturday I come in, there is still a hole. Just not as big.

We have an uneven number of two year olds in our barn. They are all getting close to their first race. The extra horse falls to me to school to the gate for the first time.

Stolen Ruler is a well bred youngster but is more nervous than the others. When I get to the gate with him for the first time, I am joined by two other first timers from another barn.

Ruler loads without incident although I can feel how tense he is. When the next horse is loaded, two stalls away, it does not sit well with my horse. His tension is building.

When they put the other two year old in the stall between us, I feel him about to flip over. I am becoming more experienced and I quickly bail off, grabbing the bars to hoist myself to safety.

Instead of rearing, Ruler sinks to his knees.

Assistants come quickly to aid us. They back the other two out of the gate, then open up the front of Ruler's stall to coax him up. I hear one of the guys say incredulously, "He didn't even struggle. He just laid down."

He thinks the horse fainted.

Once the other two horses are gone, we are reloaded by ourselves. Ruler pops quickly out when they open the gate.

Next day, I come down with Cathy. When the gate opens, I tear out a full length ahead of Cathy. By the time the gate crew adds the bell the following day, Ruler launches himself out at full bore.

He is never beaten out of the gate.

Chapter Fifteen

FORT DREARY

Doc has not won a race in a couple of years. It is a wonder he is able to get any stalls at all at the **A meet.** When May arrives, Doc is offered some stalls at Woodbine and the rest at Fort Erie.

This puts Doc in a bit of a quandary. He approaches me and asks if I would be willing to run the barn at Fort Erie. He will hire some people to help me and come down every other day to supervise.

I am not so sure if this is a good idea. Doc recently cornered me in a stall and demanded I kiss him. So this is what Mr Annesley and others around the track know. Doc chases skirts. Although I refuse to kiss and order him out of the stall, relations between us have been frosty ever since.

I also do not have fond memories of Fort Erie For me, the place is full of bad experiences and I always end up spending more money than I make. However I hear the Boyfriend is staying at Woodbine so I will not have him to contend with.

My sister Anna is back at Woodbine for a second summer, walking hots for Gordon again. It helps me enormously to have someone besides the Boyfriend to share my time off with. It would be fun to do things with her over the summer.

Yet running the barn at Fort Erie is such a good opportunity for me. After thinking it over, I take Doc up on his offer. Then the Boyfriend hails me to come by his tack room. He announces he has changed his mind and decides to go to Fort Erie, too. I will just have to deal with him.

Cathy is staying behind. She works for Doc at his farm as well as galloping at Woodbine. We go out to dinner one night and she shares her woes with me. Doc has made more than one pass at her which she does not have the skills to deal with. Cathy's boyfriend lives hours away and Cathy has been pouring out her troubles to him over the phone. Shortly after I leave, he proposes marriage and she accepts. I never think to ask for her contact information to be able to stay connected with her. I am sorry when we lose touch.

I leave with the new fillies and a colt that just shipped in.

Some one has taken the Jockey Club to the Human Rights Commission and won for us, the females, the right to live on the backstretch. This is a huge financial relief to me. These rooms are free although there are no cooking facilities. The other advantage to living on the backstretch is there is always friends about to visit with.

I decide to keep a room at the dorm, to have a home base. I move my stuff in with my friend Sandy. She can be relied on to look after my things.

Fort Erie is located directly across the border from Buffalo and depends on drawing the Americans over for the betting. The town has one coffee shop, half a dozen bars and a few more family restaurants. It has no culture, no shopping, not even a movie theatre. It is just a tourist town. I think its monicker Fort Dreary is very suitable.

In addition to galloping, the Boyfriend picks up a job as an outrider for the races in the afternoons. He is renting a room in a house nearby. I also learn that Jaylee has moved down to Fort Erie and is staying at a hotel. I am uneasy about this development. I am unsure what the relationship is between Jaylee and the Boyfriend.

The Boyfriend and I continue to have an on again, off again relationship. He is working in a barn just two rows over from me which

I have to walk through on my way to the kitchen. I feel distinctly uncomfortable going through his barn. I cannot fathom why the people there seem to hate me. It never occurs to me that he is telling tales about me.

I do not have access to a television. When the Kentucky Derby is run on the first Saturday in May, I learn that it is won by a filly. In fact, it is Genuine Risk, the very filly the Boyfriend touted last year. I am impressed.

Doc hires two grooms for me. Vicky is sixteen with no experience of any kind, Jane is twice her age and has ridden quite a bit. It is up to me to teach them the basics of their jobs. I do most of the galloping with Normie riding a couple for me. Normie is a young, polite, good looking jockey and Vicky falls hard for him. Whenever Vicky spies the jockey in our barn, the young girl dashes into the nearest stall to give her hair a quick brush and make herself more presentable.

I give Chance, the filly that kept bucking Cathy off, to Normie to ride. I gallop the filly only once and quickly learn if you put your leg on her, she either bucks or slams to a stop. This must be why chance kept dumping Cathy. She gallops beautifully for Normie and we ride together in a set. Sometimes Normie will take her back and come up on the other side of me, teaching Chance how to **rate** and come abreast of another horse.

While Jane is married and lives a short distance away, Vicky moves into a tack room adjacent to me. After the first night, she complains she was up all night fending off bugs. I assume she is exaggerating. However, in a couple of days, her wrists and ankles swell up and I have to take her to a doctor. The grizzled old physician takes one look at her and pronounces, "Bed bugs."

I drag Vicky into the racing office to show the officials her allergic reaction. They move us to different rooms and spray the others, which solves the problem. Until now, I didn't know that bed bugs were real.

The Boyfriend continues to play me for a fool. He twists my emotions around until I am completely unsure of myself. The more

I ignore him and try to walk away, the more he chases me. Yet I cannot seem to break his hold over me. When do we fight, I am giving as good as I get. I cannot make head-nor-tail of my life. I am a yoyo being jerked around, on a rollercoaster out of control. My behaviour has become more erratic and he uses this as further proof of my inability to get along with anyone. I have a lot of growing up yet to do.

When an acquaintance asks me if I follow a particular sitcom on television, I reply, "Watch a sitcom? My life IS a sitcom!"

The Boyfriend comes by my barn, promising to take me out for supper. When does not show up, I drive over to where he is staying to find him eating at his host's barbeque. While he sees me in my car, he ignores me. Furious at being had once more, I deliberately goad him by bumping his car with my own. THAT gets a reaction. I can tell by his quick movements and red face how angry he is. He flies out to the street and tries to reach through my window, possibly to strangle me. I roll up the window and drive away. I can hear him screaming behind me.

Our fights are escalating. At the rate we are going, one of us is going to kill the other. My emotions are so constantly jerked around, I have lost any self control I had.

Next day, the Boyfriend suddenly appears at my tack room. "The stewards want to see both of us in the office, right now," he barks at me. Obediently, I follow him into the barn.

I never have a chance. In the centre of the shedrow, away from anyone's view, he wheels round and hits me without warning. The first blow knocks me to the ground. "You cow!" he screams. "Take that!" I lose track of how many times he hits me. He stomps on my head. I cannot get up; I cannot get away. Now I am screaming for my life, fearful of how much he will injure me. Finally, I hear someone yell his name and he stops.

I am bleeding freely from my nose. Emotions tumble through me – anger, fear, hurt that he could do this to me. I need to be safe. In an instant of clarity, I know what to do. Grabbing a towel from my room to staunch the bleeding, I march straight into the racing office. The shocked officials reel at my appearance and ask who has done this to me. I name the Boyfriend. Since he works afternoons for the track, he is well known to them.

As I sit in an inner office, I hear the exchange in the next room. The Boyfriend is immediately expelled from the track. He has lost both of his jobs. I am safe, at least while I am on the track.

I also go to the police and lay charges. The police are reluctant to help me because they do not believe I will actually carry out my threat. I am determined to end my problem for good.

It is a huge relief in not having to see the Boyfriend on a daily basis.

He is not the only challenge I have. One morning just as I am heading home after galloping Miss Beckan, an exercise rider enters the track and begins to gallop down the outside rail straight at us. The other horse collides with us and my filly rears straight up. I am dumped over her shoulder, landing head first, with my right foot still caught in the stirrup. In an instant, I recognize how dangerous my position is and am terrified. If the filly runs off dragging me by the stirrup, I will probably be killed. The people standing at the in gate are white faced with fear, unable to help me. Even the outrider can do nothing for me. To chase my horse will only cause her to run faster.

Still upside down, I have the presence of mind to save myself. In the space of a couple of seconds, I kick my right foot free with my left.

As the filly runs off, I crawl to my feet and hang over the rail. I felt sick to my stomach and had a massive headache. Concerned people rush to me. There isn't much they can do.

The outrider catches the filly and I lead her back to the barn. I hand her off to the girls and give them instructions to finish up without me. I promptly crawl into bed and stay there until the next morning, nursing my concussion.

Anna hears of my accident through the grapevine and comes down to see that I am alright. One good thing about the track is the speed at which the grapevine operates. I had no way of contacting her to let her know I had been hurt.

Next morning, I learn the great jockey Avelino Gomez died the previous day in the Canadian Oaks. The horse he was on broke down, pitching him onto the track where he was immediately run over by two other horses. The ambulance picked him up and ferried him to the local hospital. Gomey joked with staff as he signed the consent form for the necessary operation. He died on the operating table.

Although he was a legend in the business, I never met Gomez. In a place where most people develop colourful personalities, Gomez was neon. He was one of our most successful jockeys, having won the Queen's Plate four times and is considered one of the great jockeys of all time, not just at Woodbine. The shock at the track is palpable. People discuss it in lowered voices. I do not know if many race trackers attend his funeral. The track tends to live for today and simply accepts death as a part of our existence.

We become survivors on the racetrack where we witness such rawness of Life. Accidents and pain are part of our everyday world. Our tough outer shell hides our softer side, like the crispy covering of a burnt marshmallow which holds the gooey goodness inside.

Doc comes down to Fort Erie like clockwork every second day to oversee my work and he leans on me fairly hard. Finally Doc tells me why he is being so tough on me.

The previous year, he also had to split the barn between Fort Erie and Woodbine. With no oversight from Doc, my predecessor had not bothered to train the horses. Doc is not about to repeat the same mistake. When the Americans next to me remark to Doc how I work just as hard when he is not around, my trainer backs off and we get along fine.

All of the horses are coming along fine except one. I cannot get Irish fit. She will gallop lovely for about a mile, then promptly run out of steam. Not only does she not lose weight, her udder actually bags up (fills with milk). Although she is only two, is it possible that she is pregnant?

We test her and she comes up negative. Having ruled out pregnancy, it has to be something else. We switch her bedding from

Fort Dreary

straw to peat moss and cut her hay back to one **flake** twice a day. Now we can control the portions she is eating. Finally, we figure she must be swallowing her tongue and we put a tongue tie on her.

The horse doesn't actually swallow her tongue. What she does is retract the tongue against her soft palate which cuts off her breathing. In essence, she is running on one breath of air.

The tongue tie is a soft strip of cloth. It is wrapped once around the tongue, then tied under the jaw. This prevents the horse from retracting its tongue.

With Irish, it works. Soon she begins to get fit and turn in decent work times.

Another filly in my barn is April, one of those quiet horses that is easy to forget. She looks like some kid's riding pony. We run her as part of an entry with Irish. Doc decides to only hire one pony boy for the post parade and doesn't bother with one for April.

Credit: *Ontario Jockey Club*

Fort Erie Grandstand

While the other horses in the post parade dance along side their ponies, April plods quietly, just as she did coming into the paddock. Obviously, she is not impressing many people because the betting is low on our entry. She wins handily by four lengths, then strolls to the test barn as relaxed as before. She is the most laid back racehorse I have seen.

Storm Sky ships in to us with a huge lump of scar tissue on his back leg. It is the size of a baseball, the result of a neglected injury. Despite being a stallion, he is a real gentleman. His owner has been conditioning him on her farm and the horse is already halfway fit.

The track vets quickly learn of Stormy's scar. When checking his identity, instead of examining his lip tattoo, they simply glance at the scar on his leg. It was more permanent than any tattoo anyway.

While the scar tissue never causes any problems, the colt has another issue. His 'crown jewels' (male appendages) are too big. When the horse tries to work, they are in the way and he will not put out his best effort.

I run the problem past my girlfriend Sandy, who has more experience than I in running horses. She happens to be at Fort Erie on the day we are running Stormy. Half an hour before I am due in the paddock, we apply some Absorbine to his genitals. He immediately tucks them up out of the way.

Stormy runs a credible race. Doc chalks it up to the change of jockeys and I let Doc believe this. Doc uses the same jockey for the next race. Again the horse runs well. Of course, the jockey had little to do with the improvement in performance.

After that, Doc **cuts** (castrates) the colt.

Within weeks of the Boyfriend being banned, I notice the frosty attitude on the backstretch changing toward me. One of the women in the barn where the Boyfriend was working tells me some of the lies he said of me. This is a new experience for me, to have someone turn others against me. I am to learn of other lies over the years.

Fortunately, the horses themselves are a tonic for my bruised spirit. Horses have a healing quality that attracts many of us to the backstretch. I have come out of the abuse a stronger person. With no one pushing my buttons, I begin to act more rationally. I relax and can enjoy my work.

What I have not counted on is the psychic connection I have forged with the Boyfriend. At night, I can feel him looking for me. It is like having a hook in my side and he is reeling me in. I do not even feel safe on the track. I suspect he will climb the fence and come looking for me.

Restless, I get in my car and drive, meandering down back roads to get away. Fort Erie is such a small town that I avoid it completely when I drive. Anywhere but where the Boyfriend is likely to look. It is late when I return to my bed.

The night before the court date is the worst. I feel insane, trying to disobey this compelling compunction. I drive for two hours, knowing he is looking for me.

Finally, I return to Fort Erie. I just have to stop and buy milk for my breakfast.

He finds me there, at the convenience store. He hails me and spinelessly I go to him. I climb into his car.

"You've ruined my life. I can't get a job on any track now. I haven't any money left." We talk for a long, long time. Finally, I take pity and agree to drop the charges.

Next morning, I stand in front of the judge, telling the learned man that I am dropping all charges. "Are you sure?" he asks. I am let go.

Outside of the courtroom, the Boyfriend hugs me tightly and gives me a kiss. I have to hurry back to the track so we say our good byes.

That night I see his car at Jaylee's hotel room. Somehow, I am not surprised. He has lied to me again. At least now he is out of my life. He is still banned from the track.

When the A meet moves to Fort Erie, stalls become available at Woodbine. Our barn is reunited. Since Cathy is gone and I do not care for her replacement, I soon quit. I have learned a lot this summer, about horses, about people and especially about myself.

Chapter Sixteen

THE LETTER

Intuitive decisions fly in the face of rational logic. They cannot be explained. They cannot be justified. They just are to be followed.

I decide to travel to the west coast where I have an older sister. I can stay with her and find work out there. I want to be away from the gossip, the ill feeling that has been generated about me all summer.

There is another reason, one that I cannot even admit to myself. I am pretty sure that the Boyfriend has likely gone to the west coast. He had mentioned wanting to try working there. Even now, I have a sense of something unfinished with him. I cannot even articulate what it is. I just have to go.

I pack my car with camping gear and set off. I pick up hitchhikers along the way to break the monotony of the long drive. I stop at the racetrack while in Winnipeg. A friend had given me a contact there and I would like to see that track.

The backstretch is tiny. I run into one of my old comrades from Bobby Fisher working there. Lee tells me Fourfoot is now a jockey and even has a ride in a race this afternoon. I wish I had time to stay and look Fourfoot up but I have a long way to drive yet. It takes me a full week to drive to the west coast.

By now, I am used to living impulsively and landing on my feet. When I go to Hastings Park, the thoroughbred track located in downtown Vancouver, I am immediately hired as a groom. I also pick up a job off the track grooming some yearlings ready for the sale. As long as I prepare them daily, I can be flexible with their schedule.

The trainer at the track I am rubbing for is up from the States. All three of his horses have a bad case of **ringworm**. This is a common fungal infection, easily passed from animal to animal by brushes, tack or any other shared item. In Ontario, we used a green bacterial soap which we left on for half an hour then rinsed off. It kills the fungus in a few days. I go to the tack store to buy it.

Now, the oddest series of coincidences unfolds.

While I am in the tack store, I over hear a conversation. It is about someone working on a farm locally who has the same unusual name as the Boyfriend. It has to be him. The farm belongs to the owner of the tack store. I say nothing, pretending not to hear the conversation.

When I go to the office to obtain my license, I see a notice pinned on their bulletin board. It is from the Jockey Club in Ontario, banning a certain person from all tracks in Canada. The person is the Boyfriend. He was actually working on the track here when this notice came through. My actions have reached all the way here.

A couple of days later, my sister (who I am staying with) gives me instructions to meet her where she keeps her own riding horse. I take a wrong turn and end up pulling into a horse farm to turn around.

Just a week after I start at the track, I am paged to the office for a message. There is a phone number but no name. At lunch, I dial the local number. The call is answered by the Boyfriend. He saw my car in the parking lot this morning when dropping someone off. I am astonished to learn he is living on the very farm where I turned around when I was lost. My psychic bond pulled me twenty five hundred miles across Canada to land right on his doorstep.

The Letter

I agree to meet the Boyfriend. He confesses to being delighted to see me again. I am wary, cautious. I learn that Jaylee has gone back to her home in the States. He does not have anything nice to say about her and seems relieved that she is gone.

We begin seeing each other again but not on a daily basis, nor are we intimate. I decide to take things slowly, find out if I can trust him. I certainly do not want to be dependent on him again.

I have been to British Columbia before. Seven years ago I spent the summer working on a small thoroughbred farm half an hour out of Vancouver. I had such a terrific boss and drive out to see if I can find him.

The area is much more built up and I have trouble finding the farm. To my disappointment, the farm has been sold. It is now a Quarter Horse breeding operation and no one knows anything about my former boss.

My new boss at the track, however, is impressed when the ringworm that he has been struggling to contain clears up within a week. He is only up here for three weeks but is very effusive in his praise when he leaves.

When my trainer at the track leaves, I find another job immediately. This new boss is a big cowboy type who wants to be overly friendly. We drive to a local tack store where he buys me a gift. Later, when he drops a familiar hand on my shoulder, I leap away from him. It is obvious what he wants.

The Boyfriend asks me to pick up his mail, which is still being delivered to the racetrack. The second time I do so, I see a handwritten letter with an American stamp on it. It is from Jaylee.

He told me they were through! My emotions are boiling. I am so angry I can hardly finish my work. What shall I do?

I drive back to my sister's place when I finish work. I know I will be alone for a while. The Boyfriend is due to come along later this afternoon. In the privacy of the house, I decide to steam the letter open. I know this is both illegal and immoral but I absolutely must know the contents of that envelope.

It is a seven page letter from Jaylee. They have not broken up. In fact, she is pregnant. He must have phoned her to tell her I am here. She is furious, calling me That Pig and other nasty epithets.

I have only read a page and a half when the Boyfriend shows up early. I meet him at the door as he cheerily greets me. I open the door a crack, hand him the letter, slam the door in his face and lock it.

As he contemplates the opened letter in his hand, he understands what I have done. Now he cannot continue his charade with me. Instantly, his tone changes. Enraged now, he screams at me for opening his mail.

I hide on the other side of the door where he cannot see me, shaking. He obviously intended me to find this letter, otherwise he would have asked someone else to get his mail. He knew that Jaylee would write.

After a few minutes, I hear his car start up. I carefully peek out the window to ascertain he is gone. Then I quickly jump in my own car. He may come back, bent on revenge and I do not want to be here if he does return.

It takes me all afternoon to simmer down. He has played me for a fool for a long time. In Ontario, no one would tell me to my face what he was up to. I now have the absolute proof that I have been looking for. These two are obviously planning a future together.

I wait until I know my sister is home from work before I return to the house. That night, I tell my sympathetic sister all that has transpired. I make plans to drive back to Ontario immediately. I have enough money in my pocket since I have already finished the yearlings and been paid for them. I certainly have no desire to be in the Boyfriend's sphere of influence.

I retrace my route across Canada. Driving gives me time to think, something I do not want to do just now. My thoughts are muddled, disjointed. I feel as if I have my tail between my legs, disgraced. I also feel unsafe somehow. I drive home in half the time it took to drive out, as if pursued by some unknown ghost, hardly sleeping until I reach Ontario.

I have spent years following my heart, which opened my intuitive nature. I have made a fatal mistake, believing that anyone with intuitive gifts has high motives and is an evolved soul. How could I have been so wrong?

At least now I have some of the answers I have been looking for. I have to put my life back together without the chaos he created. I need to understand how I could have been so wrong.

Chapter Seventeen

WINTER OF REFLECTION

"Hey, Jan, what are you doing these days?" Liz Ashton greets me warmly from her office.

"Actually, I'm here looking for a job." Here I am back in the Humber College Equine Centre, where I took my horsemanship course three years ago. The school posts job opportunities on a bulletin board in the classroom. I thought this might be a good place to start looking for a winter job.

"What are you looking for?" Liz asks as she pushes the door wider to let me into her office.

"Just something for the winter, until I go back to Woodbine in the spring."

"I'm looking for someone to work at my farm over the winter. Are you interested?"

"Sure."

Presently, Liz is the director of the horsemanship program. She taught me stable management in my first year, then took a sabbatical the following year to compete in the Olympics. She shows her own horses during the summer but in the winter, she takes on other horses to train, including some youngsters to break for the track. I know Liz will treat me well and decide to take the job.

Her farm is located some forty minutes due north of the track on the edge of ski country. It is quite small compared to the fancy farms I have worked on. There are only three employees. Jennifer manages the day to day operations and sets such a fast pace that I can barely keep up with her. A graduate of Humber as well, she runs the main barn, where I spend most of my time. Sue is a working student and she has the smaller barn to run. She keeps one of her own horses in that barn plus gets a lesson from Liz every week in lieu of wages. I simply don't know where she finds the energy to ride after work. The three of us inhabit the main floor of a house with Liz living upstairs. I share a bedroom with Sue.

I am in awe of my coworkers. This is a new experience for me to work in a show barn. I find the routine a bit overblown. Whiskers on all horses are trimmed off every three weeks, even the racehorses. Feet are oiled daily as part of the grooming. For the first time, I have to deal with switching blankets. By midwinter, each of the show horses are clipped, then wears three blankets, piled one on top of another. To go outside, those blankets are stripped off to be replaced by three other blankets. When the horses come in, the process is reversed. It is a tedious, time consuming chore.

I am impressed with Liz's dedication to her sport. By 6 am, when I first wake up in the morning, the lights are already on in the arena. Liz is out there jogging. Then she rides one or two horses before heading off to teach at the college. She cleans her own tack when she returns from work, forbidding us to add that to our workload.

Horse people are always drawn to the other animals around them. The more permanent set up of the farm allows me a chance to befriend the three wonderful pets housed there. They are a delightful diversion from my work.

There are two orange cats named Reddy and Ruffy. Reddy is a simply striped feline and a mouser par excellent. Every time you see her, she has a fresh kill in her mouth. She catches an average of six mice per day. Ruffy, on the other hand, is more leisurely. She is gorgeous, with a generous white bib down his front. She will catch only a single mouse per day, then take it round to each of us in turn to show us how clever she is. She brags incessantly, dropping the live mouse and pouncing again to show us how she caught it. Often she then loses the mouse behind a tack box and then spends another hour catching the mouse again. Once you praise Ruffy lavishly, she heads off to show her mouse to the others. Poor Reddy feels so inferior. She starts bringing the mice to show us that she, too, can catch mice.

Snydly, Liz's dog, is of mixed breeding but mostly German shepherd. Certainly he is one of the friendliest canines I have encountered in a long time. For the first time since my childhood doggy pal was put down, I fall in love with a dog. Snydly starts each morning by seeking out each of us in turn for a greeting. He gives you the feeling you are the most important person on earth at that moment. Nothing doing but you must put down your pitchfork, squat down and give him a real hero's welcome. When the mutual greeting is finished, he moves on to bestow his blessings onto the next human-god.

Liz takes her dog with her to the horse shows, where Snydly has perfected the art of begging for food. With a twist of his head, thumping tail and adoring eyes, who can resist? Unfortunately, in the evening after the show, Snydly routinely comes down to our apartment to throw up on the rug, much to Liz's embarrassment. Finally she is forced to hang a sign around the luckless dog's neck at shows 'Please do not feed me'. I have visions of Snydly sitting with the silly sign on, twisting his head, begging away. However, the dog no longer gets sick after a show.

Liz comes down the stairs one night to ask me a question. "Do you know of anyone who would be interested in a job for the winter in the States? I have a friend in New York looking for a girl."

I think immediately of Sandy and recommend her. She flies south and becomes a valued member of their staff for the winter. Liz lets me know how pleased the Americans are with Sandy.

Sandy, who was born in Britain, flies home that winter, to attend her citizenship hearing. The irony is not lost on me, that she was born in England and is working in the States all while applying to live in Canada.

Everyone at the farm is incredibly nice to me. The girls invite me out shopping and to watch T.V. but I feel a reluctance to join them. After the Boyfriend, I am still wary of friendly advances. I cannot trust enough to confide my pain to these two strangers.

Night after night, I dream of the Boyfriend. Even in my dreams he rejects me. I just can't seem to let him go. I miss him fiercely. This is illogical because he never loved me, didn't want me and treated me like dirt. I am consumed with thoughts as to why I was never good enough. Is it possible to love and hate a person at the same time? I don't understand why I believed his lies for so long.

My soul is raw. My emotions are in free fall. Depressed, I gain weight like crazy. I pack on the thirty pounds I lost my first summer at the track. I burn his old love letters and pour my soul into my journal, grieving for a future that can never be. I am trying to make sense of the invisible chord that binds me to him.

During the day, I seem easy going and friendly towards my co-workers and then the sun sets and my mask slips off. I need to get away and leave almost every night to see my friends from the track to talk. And talk. And talk. I am scattered and shattered, trying to put my life together again. I swear I will never go down this road of abuse again.

My compulsive self-analysis begins to pay off. I am finally able to see the game he was playing, picking women who felt inferior, who would look up to him yet feel sorry for him and want to help. He knew how to push my buttons so I would recreate problems, always be wrong and stay dependent on him. Still, why wasn't I able to walk away sooner, when I knew all this? Answers come late one night while talking with Rhonda. In my mind's eye, a series of pictures appear. As I focus on these pictures, they prove to be memories from other lives. I have had past life memories before, which bubbled up under hypnosis or in vividly emotional dreams. It takes me days to unravel these past life memories that have just surfaced.

I have been connected with the Boyfriend lifetime after lifetime. We have alternated being victim and perpetrator, our karmas binding us together. Now that I see my own part in all this, I am less a victim and more responsible for my own actions. I can step off that wild roller coaster of emotions that has enslaved me for three years and many more lifetimes. I am more centred, able to think more clearly.

When I meet other women who cannot walk away from abusive relationships, they tell me, "I felt like I had known him forever." I understand now where they are coming from. This is how I feel, too. I decide to break away from him without vindictiveness, hoping this will uncouple the chain that has bound us for so long.

With farm work being pretty mundane, this gives me plenty of time for thinking. I decide the best solution is to be wildly successful. I optimistically dream great goals and imagine meeting the Boyfriend in the future, modelling how he did not succeed in ruining my life. I plan to lose weight and become an exercise rider, also to work in Kentucky and England. I will not be stymied by his manipulations but rise above them!

Firstly, I need to get back riding. I learn that I can take a riding night course at Humber to finish my diploma in Horsemanship. I had some setbacks while at school and was not able to finish my last semester. I re-enrolled but the College overfilled the program and they denied me the opportunity to ride. There simply was not room. When I drive to the school to register for the night course I need, I discover that the class is already full. Then the registrar mentions there is a new, higher level being offered. Feigning a confidence I do not feel, I mention working for Liz and talk my way into that course.

The course is weekly for ten weeks. While many of the wing-ding horses that I previously rode are (thankfully) gone, there are a few on our class list that I struggled unsuccessfully to ride while still a full time student. I deliberately choose them. To my delight, my skills have improved to the point that these horses are a lovely experience, slightly but successfully challenging. I notice none of the other riders manage as well as I do (I keep this to myself) but my confidence soars. By the end, I easily jump the required hunter course and pass.

At last! This course completes the credits for my Horsemanship diploma from Humber College. It seemed everything conspired against me while I was a full time student there – ill health, a badly timed operation, witnessing a tragic accident along with mostly unsympathetic teachers. Earning my diploma is a personal victory and I feel cheated when my diploma arrives uncerimoniously in the mail. I worked so damned hard for that certificate, overcoming numerous obstacles (including my own poor attitude) and feel I deserve a proper graduation!

Being on a farm does not insulate me from the heartaches of hurt animals. One of the broodmares on the farm is a rather pretty thing with an ugly knee that she cannot bend at all. The joint was damaged in racing and is fused together. The mare gets around fine, walking and cantering but she cannot trot. This puts her at the bottom of the **pecking order** but she throws beautiful foals.

One morning after my day off, as I am mucking the **run-in shed**, I find blood soaked straw. As I exclaim aloud, the two girls exchange glances. Sue admits "The foal's dead." It is the foal from that beautiful mare.

Stunned, I listen to how they found the young animal the morning before. The fence around this paddock is split rail. Playing in the night, the foal had run into the fence and impaled his chest on the broken rail. He had returned to the run-in shed, laid down on the straw and died. I had not realized these picturesque split rail fences could be so deadly. Even on such a well run farm, accidents still happen. With their tendency to panic and run, so many horses are injured before they are even broke to ride but it deeply disturbs me to lose such a beautiful young foal.

Late one afternoon, one of the horses starts to **colic.** Colic is the condition that all horsemen dread, being the biggest killer of horses. If the horse rolls, he can twist his intestines and die so he must be monitored until he recovers. The other two girls make arrangements to stay up all night with the horse.

I go to bed early. There is no point to me staying up because I may be the only one at work tomorrow. In the morning, the horse is still ill. The girls decide to ship with the horse to Guelph, where the large animal clinic is. Sure enough, I find myself with twenty five horses to feed, muck and turn out. Only the grooming can be skipped. The girls return about eleven in the morning and go straight to bed. It takes me an hour longer than usual to finish by myself and I am completely worn out. I cannot complain as the two girls worked equally hard.

Fortunately the horse recovers and comes home a couple of days later. We continue to nurse him for a few more days and life on the farm returns to normal.

Most of the riding on the farm goes to the other two girls, due to their seniority but I still get to do some riding. Now, as racing season approaches, the yearlings we started are prepared to ship to the track. There are plenty of horses to go around and I am given a couple of regular rides

Liz carefully sets up an interval training chart for us to follow each day. Interval training is a scientific way of conditioning which minimizes the chance of breaking down the animal. There are so many minutes of trotting and cantering with walk breaks, all carefully timed.

The filly I am assigned to ride is so sensitive that if I turn my hand over to read my watch, she takes off. After a few rides, I realize the solution is to fasten my watch on the side of my arm so I can read it with my thumbs still held uppermost. Strange solution but it works.

Also, I ride Arthur, a leggy 17.1 hand two year old. He is so tall that when Liz initially backed him, he crossed his legs and fell over. Talk about a lack of balance. I ride Arthur just before lunch. One Friday morning, he is hot and sweaty when I am finished. Usually, I dash into town twenty minutes away on Friday lunch to cash my cheque (this is before ATMs). If I walk him cool, I will be too late to get away.

"Just throw a cooler on him and leave him in his stall," the girls advise.

When I come back from lunch to check on Arthur, I am horrified to find he has ripped the expensive cooler into confetti. There isn't a piece of the blanket bigger than six inches square. What a mistake to leave the cooler on him. I feel so guilty for ruining the cooler even though the girls suggested it.

My day does not get better. For some unremembered reason, I am alone that afternoon to finish chores. Everything conspires against me and I get later and later. I am already in a foul mood due to the ruined cooler.

A foal is locked in the run-in shed while recovering from an injury. To give the foal some exercise, I have to bring other horses in from the paddock, then turn the youngster out by himself. While trying to get the foal out of the shed, he pulls loose from me, still with the shank attached to the halter.

Now I have to catch the rotten little bugger in order to get the shank off. Fifteen minutes later, he is having no part of being caught so soon. Finally I lose my temper. I indulge in a supreme fit of anger, leaping up and down while screaming at the top of my lungs. What I do not know is that Liz is entertaining friends and they are watching me. In fact, I have become the afternoon's entertainment. It takes me another ten minutes to catch the foal. I still have to muck its stall and bring him in. It's a very late day when I finish.

With the race horses due to ship into the track, the date of my departure is set. Liz comes down from her apartment to talk briefly to me before I leave. "You know, this is the best crew I have ever had at the farm." This is encouraging for me to hear. I was pretty sour on work when I started at the farm. With my inner wounds healing, I am ready to face the track again.

Chapter Eighteen

WORKING WITH SANDY AT GREENWOOD

Sandy is stunned by my weight gain when she returns from the U.S. My face is round, my thighs are wide. I have tried to get a galloping job but can find no one who is willing to hire a chubby girl with limited galloping experience. Instead, Sandy gets me a job rubbing for her boss. Mark, the trainer, is a middle aged, intensely private man. He has a couple of loyal owners, gallops most of his racehorses and wins a decent percentage of his starts. There are a few nice horses in the barn. I am given a string of horses to rub. Mark also lets me get on horses here and there. A hotwalker rounds out our crew.

Sandy is a wiry strawberry blonde with no higher ambition than to look after her horses. She started at the track working for Mark and has never worked for anyone else. As head groom, she is knowledgeable and totally dedicated. She has a lot to teach me. I suggest to Sandy that we rent an apartment together. Having done the math, I know that what we will save on cooking our own meals will offset the extra money for rent. She enthusiastically endorses my suggestion. She doesn't have a car and she loans me the money for rent deposit. We find an apartment just minutes from Woodbine.

While it might seem a bit close to live and work with the same person, Sandy is the opposite of the Boyfriend. She is quiet and reliable, a real friend. After the chaos of the past three years, it is wonderful to settle down. I relax away from the prying eyes of the racetrack. I also enjoy feathering my nest, even if the furniture is scrounged. I decorate, cook and we entertain our friends.

As soon as Greenwood opens, Mark moves the barn down there. I have never been based at that track before. Greenwood is much older and smaller than Woodbine and the racing there alternates between the Thoroughbreds (flat racing) and the Standardbreds (harness racing). With just one ¾ mile galloping track, it is used for both training and racing. On the north side is the grandstand where the trolleys from downtown dump eager betters on race days. On the south side, a dozen barns lie adjacent to the track with parking lots bookending the track. The track kitchen is situated half way down the backstretch with the test barn tucked in behind. Between the barns are cement pits into which we muck. Almost every inch of the backstretch is used for something. There are no hot walking machines and precious little grass. The barns are much wider, twice that of Woodbine with a generous shedrow and stalls on both sides. The barns are fully enclosed and are toasty warm in winter.

The drive down to Greenwood takes us half an hour. The sun is just rising, traffic is sporadic and the sky is streaked with pink and baby blue. Most of the city still sleeps. The peaceful mornings start the day off softly.

After the chaos of the past couple of years, it is comforting to work uneventfully, to let each day flow as gently as the day before. The days melt together and I enjoy working at the track thoroughly.

It is so easy to work with Sandy. There is always a radio playing and I sing happily to myself as I muck stalls. I crank the radio up for my favourite songs; Mark turns the volume down again as soon as he notices. We play this game over and over again each morning.

Sandy's and my stalls are on either side of the aisle and Mark alternates galloping a horse for each of us. With the barn close by the track, there is often time enough to slip out and watch a horse gallop.

If we are not racing, Sandy and I drive home about noon. We alternate feeding in the afternoon. This way, each of us gets two or three afternoons off per week. After the continuous work in the past two years, this is heavenly. However, another fly in the ointment is the change in racing hours. The Jockey Club has now introduced twilight and night racing to boost attendance. Twilight racing starts at 4 pm and is finished by 7 o'clock. It does not add much time to our day.

The same cannot be said of night racing. It begins at seven and ends by ten. However, the groom has to work from 6 a.m. until 11 a.m, be in the barn to jog the horse that is racing for soundness for the **Commission** vets (sometime between two and three o'clock), do afternoon feeding at four o'clock, then racing starts at seven. By the time you cool out your horse from a late race and do it up, the time we finish is closer to 11:30. Now our race day stretches almost eighteen hours, yet we are not paid a dime more, and in addition still expected to be back on the shedrow next morning by six.

While the change in the racing schedule is a godsend for the betting public, it has the effect of draining the backstretch of much of the best help over the next few years.

Sandy, as senior groom, has the best horses to rub. I rub the unproven two year olds and unproductive older horses. Veracee is my problem child. She is a skinny three year old maiden with an equally long narrow face accentuated by a vertical stripe. This earns her the nickname of Skinny Face from the other exercise riders in the barn. Besides her natural build, the chestnut filly is a **poor doer**. She likes to dunk each mouthful of her hay in her water pail as she eats. When the water pail is dirty with hay, she just quits eating. No wonder she is underweight. She is also tough to gallop, tough to settle. Anything can spook her. The tractors working on the turf cause her to brake in mid stride and pull a U turn before Mark can even react. She drops Mark so often that he finally brings his Western saddle in from the farm and puts that on her. Mark looks pretty funny riding in that Western monstrosity alongside the other exercise riders standing so daintily in the tiny saddles. Veracee is the least successful horse in the barn. She is running dead last for $6250, the lowest priced claiming race on the A circuit. She doesn't even try, loping around with the pack without putting out any real effort. My job is to improve her performance.

My second horse to rub is Mount Herman, a giant of a two year old colt, standing 16.2 hands high. This bay colt is not blessed with an abundance of brains so I nickname him Dum-dum. I am told that as a yearling he jumped a five foot gate from a stand still, clearing it without leaving a scratch on himself. He doesn't show any real talent for running, however.

My dislike of Tri Fran lasts until the first time Mark lets me gallop the colt.

I also rub Tri Fran, a well bred, good looking two year old colt. With the stud fee for his sire set at $15,000.00, good things are expected of this young horse. I am not impressed. I find him annoying to look after - 'coltish'. He is always nipping at me, challenging my space. The total opposite of a sweet filly like Vera. My dislike of Tri Fran lasts until the first time Mark lets me gallop the colt. As I ride back into the barn, I remark to my boss, "Mark, it's love." Indeed, I have totally fallen under the spell of this well-balanced and athletic horse.

Horses are as unlike as people are. If Herman can be compared to a dump truck, Vera is like a cheap foreign car you can't rely on. Tri Fran feels like a Rolls Royce. Tri Fran is a good feeling devil to walk. He flies around the corner of the shedrow and leaps into the air two or three times. He finishes with a buck and a proud snort. This behaviour makes me laugh but I worry about getting him to the paddock without turning him loose on race day.

One of our horses has developed a problem. The vet arrives to Xray the ankle. While Sandy suits up in the lead filled apron, our hotwalker Jerry hovers over us, curious to what is going on. I am on the end of the lead shank.

"Whatsamatter with him?" Jerry nosily asks.

Of course, if we knew what the problem is, we would not be Xraying.

"Well, now," answers our young vet, "I don't like to swear in front of the ladies."

Swear? I think to myself.

"... But it's fucked."

I relate this incident to my parents a few days later. Even in our house, which has a strict No Swearing policy, Mom and Dad howl with laughter.

Sandy's **big horse** is a hyper but talented gelding named Karel Van Kaat. The 'big horse' in the barn has nothing to do with his stature; the expression means the horse that is the most valuable, the money earner. As a two year old, Karl had distinguished himself enough that he is now thought of as a possible **Plate** horse. I think the horse is border-line crazy but Sandy loves the horse.

The care Sandy gives this horse is exemplary. As soon as Sandy arrives at the barn, the horse is already wound up. She throws a shank on Karl and gets him walking or has him galloped first. Then he is cooled out and put in his stall with fresh hay. By late morning, she walks again, usually for another half hour. Actually, he doesn't walk. He jogs the whole time he is led. By feed time, Karl is so wound up that he needs to be taken out again and hand walked, usually by Sandy herself. The fretful horse relaxes in her care, keeps his weight on and runs like the fool he is. He wins half the races he is entered in.

After another impressive win, there is an offer made to buy Karl. Mark and his brother Mike, who jointly own the horse, talk the offer over. They want $100,000 for the horse but settle for $75,000 and a percentage of his future earnings.

The horse never makes a dime for his new owners. Not every groom has Sandy's dedication.

When we race at night, Sandy usually sleeps in the tack room during the afternoon while I nap in my car. When the racing starts, we each hole up in the tack room with a book until our race is called.

It is quiet in the barn at night with just the sound of horses eating their hay. An occasional hotwalker swings by with a horse wrapped in a cooler. This night, Sandy and I sit in the tack room with the door open so she can keep an eye on Bo, standing in ice water up to his knees in a plastic tub. This is a time honoured tactic to help sore horses run better. As a six year old, the horse is fairly reliable at staying in his tub. He amuses himself by playing with the shank dangling from his halter as he is not actually tied up.

From time to time, some announcements from the P.A. punctuate the quiet. These calls mark our progress through the night. Results of the fifth race. Which horses are called to the test barn. First call ("Get your horses ready") for the sixth race. Second call ("Bring your horses along").

There is a comfortable monotony in these announcements. Without looking up from my book, I mumble, "I thought you were in the sixth."

There's a huge gasp from Sandy. "I'm so used to being in the seventh, I completely forgot which race we're in. JERRY!"

Our barn is farthest from where we get our **head numbers** before crossing over to the paddock. Sandy should have started out ten minutes ago. Not only do we have to get the horse out of the ice tub and have his pee, but Bo is so headshy that the bridle must be taken apart to get it on him. Then, because Bo is a stallion, the halter must be fitted over the bridle and the wrapped portion of the chain of the lead shank is fitted over his gums. The lip chain gives control because otherwise Bo would try to savage the pony. We have all these steps to take before Sandy can leave. It is a $200 fine to the trainer if you miss getting your horse to the paddock for the race.

While Jerry and I wrestle the huge tub of water out of the way in the front of the stall, Sandy whistles for Bo to pee. Whistling prompts the horse to empty his bladder, thus running better. Acting as a nursemaid, I pull the bridle apart and hand it to Sandy, then prep the halter and shank. In less than 5 minutes, Sandy is jogging down the pavement, fumbling for her cigarettes as she runs. Sandy jogs the whole way there and barely makes it to the paddock on time.

Another night, when Hussy is cooling out after a race, she is so hungry that she manages to pull Jerry over to another horse's haynet to help herself while she is being walked. She is still hot and not yet relaxed or having had her fill of water. She chokes on the hay.

Choking is a weird phenomena. Horses are just as capable as people of choking on food and the results can be just as deadly. Because horses have a bigger windpipe, usually some air gets through allowing enough time to get a vet. However, horses cannot vomit and the mucous that surrounds the bolus of food could kill them.

Sandy puts out an emergency "any vet on the grounds call over the P.A. Because it is late, most of the vets have gone home. Fortunately Smitty shows up. I have not seen Doc Smith since my college days, where he had been a lecturer. I don't think he remembers me but I am reassured by the presence of the gruff old vet. In minutes, he has the situation under control.

Sandy decides to stay the night. The tack room already has a mattress so we dig up enough coolers for her to use as blankets. I drive home, promising to bring clean clothes for the next day. Smitty, bless his heart, comes by every half hour until 2 am to check on our filly. I am reminded again of the incredible dedication to these animals by nearly everyone on the track.

Grooms rarely get credit for their horses' performances. Yet every trainer relies heavily on his grooms for the well being of the animals in their care. A good groom is worth their weight in gold. They know how to keep the horse sound, in good weight and most importantly, winning. I am learning all the tricks of the trade, how to maximize my horses' performances. I, too, am becoming a top notch groom.

Chapter Nineteen

WORKING WITH SANDY, SUMMER

In May, we ship back to Woodbine for the racing there. With the weather turning milder now, Mark decides to cut (geld) all three colts in the barn. Mark has one colt cut at a time and lets each horse heal before doing the next colt. The biggest concerns with gelding a colt are infection and swelling. To combat the swelling, forced exercise works best. This requires me to ride the boys for half an hour in the back field during feed time on top of their morning gallop. Since Sandy doesn't ride, I have to be back every afternoon to exercise the newly gelded horses. I am young enough not to be cautious. I love the freedom of riding in running shoes and do so, since there is no one to forbid it.

One afternoon, while riding Herman, he bucks and my foot slips through the stirrup. I am very, very lucky. I stay on the horse and manage to free my foot. If I had hit the ground, I could so easily have been dragged to my death. This is the last time that I will ride in running shoes. Lesson learned.

After his morning gallop, I am now required to cool Herman out with cold water bandages. Absorbent bandages are soaked in water and wrapped on his front legs as he is cooling out. They helps the legs to stay cool and not fill up with oedema. Every turn or two of the shedrow more cold water is poured over the wraps, using a Styrofoam cup.

Initially, I put the cold water in a pail just past where Dumdum's drinking water is hung. The horse waits until I lean over, then spits his drinking water onto my back. I move the cold water farther down the aisle but he holds the water in his mouth and still spits it on me. Finally, I am clever enough to reverse the order, putting the cold water before his drinking water. Would you believe that damned horse walks all the way around the shed with a mouthful of water just to spit it on me anyway?

I am genuinely worried about getting Tri Fran to the paddock for a race, since he is so tough to walk in the barn. I needn't have lost any sleep. We run Tri Fran and Sandy's two year old Long Ford as an **entry**. Long Ford is absolutely nuts. Despite the lip chain Sandy has on him, the dark horse pulls Sandy completely off her feet several times. Without the strong restraint, I suspect she would not have gotten him to the paddock. He is also silly to saddle.

Tri Fran finds this of great interest. In fact, it is so fascinating that it never occurrs to him to act up. He is only a little fresh to take over. Fran cools out just fine but Long Ford is thoroughly worked up. Three hours after we return to the barn, this horse is still lathered and full of adrenalin.

I only have so much patience. It has already been a long day. I am ready to go home but Sandy and I still share my car, as well as our apartment. Finally I threaten to leave Sandy there all night but there isn't anywhere for her to sleep here. We bathe the horse one more time and throw him his grain. Sandy whips on his wraps as he eats. This is the difference between Sandy and me. She is a born groom. She can tolerate the nonsense of a bad actor. I find the repetitive nature of grooming to be stifling. I know I cannot be a groom forever.

Horses only run well when they are feeling good. I need to get some weight on Vera to encourage her to be more full of herself. I hang a second water pail in her stall. After her lunch, I pull her water, scrub the pails and leave her with fresh water for the afternoon. At afternoon feed, I change her water again. Vera begins eating with more gusto.

I am also cooling her out myself more often. When I do, I encourage her to play about on the end of the shank. As we turn the corner where she can see horses on the track, Vera leaps about, often doing a **capriole.** Playing with her works and her whole attitude changes. She is now sassy, pinning her ears and pretending to be tough. More importantly, she begins to run.

At the moment, I am rubbing three fillies, Vera and two three year olds. With the luck of the draw, Mark enters all three of them for the same day of racing and all three draw in. The fillies are an entry in the first while Vera will run in the third.

Since Sandy has to run one of my entry, she enlists a couple of her friends to walk the fillies running in the first race. As soon as I finish with my filly in the first, I will immediately take Vera over for her race. Jerry can cool out Vera.

The two girls show up before Sandy and I leave for the paddock. When we come back, they are organized with the bath water set, the coolers ready, the drinking water hanging on the wall. They have also mucked and bedded the stalls for my two year olds and readied Vera for her race. All I have to do is bath my filly and slip the bridle on Vera. By hustling, I am right on time to start for the paddock again.

I enjoy running horses. It begins with the P.A. announcement to "Get your horse ready for the sixth race (or whatever race you are in), first call." That is the cue to start the grooming. A quick curry and a gloss over with the soft brush, followed by the rub rag to tidy everything up. The feet are picked. About then, the P.A. warns again. "Get your horses ready for the sixth race, second call." Now the halter is slipped off and the bridle buckled on. Depending on the horse, the halter may be buckled over the bridle, especially if a second groom is going to lead from the other side (not very common) or a **lip chain** is used (stallions or very difficult horses).

Now the P.A. exhorts us to "Bring them along for the sixth race." Grooms whose barns are farthest from the tunnel may have already started out. Since Vera wears blinkers, I grab them to carry with me. As I head out, inevitably someone will wish me good luck. At the tunnel, I meet the voice who is directing the flow. The official verifies which horse I have and buckles the head number onto the right side of the bridle. I move out of the way and continue to walk my horse about, spacing myself so no horse can kick us. I listen for the race before us to run. Now the grooms get the go ahead to start towards the paddock.

The tunnel is under the Marshall (or outside) turf course. On the other side, we emerge onto the main track. This is where I experience what the footing will be like for our race. A day or so after a heavy rain, the track becomes a gluey clay that can suck the boots off a luckless groom's feet. Sometimes I see a groom with only a sock on as she wrestles her boot out of the mud. We all dread a heavy track. It is tough on the horses as well as the grooms.

Today the track is fast, after days of sunny weather. I cross the track and begin walking up the black cinder path on the outside of the oval. This path absorbs the hot summer sun and lacks any shade, making it a brutal march. I pass the **meat wagon** parked here. The driver is friendly with many of us, having been a groom himself before landing such a cushy job. We insult each other and I continue on. Finally, I enter the cooler space under the grandstand. The pony boys are resting here in the shade with additional ponies tied to the wall. I turn left toward the saddling enclosure.

As I emerge back into open air, there are people pressed against the rail on both sides. They are trying to learn something about the horses that will make their betting selection easier. I ignore them, cross the outside walking ring and enter the saddling enclosure. I present Vera to an official with an open binder and twist Vera's upper lip back to expose her tattoo. He verifies her number against her Jockey Club registration paper before him. We enter our stall, where Mark and Vera's owner are waiting. The saddling enclosure can be a busy place. Owners and guests are allowed to stay close to their horse. Often one or both of Mark's brothers comes to the races. They join us in the paddock to be part of the action.

When the jockeys are released from the jocks' room, they swagger down to their respective mounts. The jockeys are immaculate in their white breeches, shiny boots and brightly coloured silks. With their chin straps dangling undone and twirling their whips, they hide their nerves. The jockeys greet the trainer and owner and nod to the groom. Final instructions are given to them. With the call, "Riders up," the trainer legs up the jockey and I start leading my filly.

As the horses start parading out of the paddock, Vera stops. The elastic girths are pinching my skinny filly. She splays her front legs like a dog and threatens to flip over backwards. I have a serious problem and I have to keep her moving. Alfie, the head outrider who leads the post parade, walks at a snail's pace. So here I am, stuck in a traffic jam with a nervous jockey, on a horse that is trying to rear with nowhere to go. I do the only thing possible. Grabbing her by the chin strap, I trot the filly in a zigzag back and forth, pulling Vera sideways whenever she tried to dig in her toes. I am barely aware of the crowd thickly lining both sides of the walkway, shouting encouragement at the various jockeys. Under the grandstand, I hand Vera off to the pony boy with the instructions to keep her going. While there is time to slip into the grandstand and place a bet, I have no interest in doing so. As the last of the racehorses steps onto the track, the bugler rides back into the paddock. With nothing better to do, I watch the post parade and the loading into the gate.

A race is the cumulation of months of work. Few experiences equal winning a race and being well beaten is a big letdown. Vera runs credibly and I am pleased. Past the finish wire, the jockeys let the horses run out of steam until the backside, turn around and gallop back. Now the jockeys are covered in sweat and dirt. The jockeys unsaddle (no one else must touch their equipment) and carry their equipment back into the jocks' room for the weigh in.

The racehorses are still revved up on adrenalin. Vera leans against me, her sand encrusted sweat biting painfully into my bare arm. As I walk down the track, I pass the horses already coming for the next race.

On this busy afternooon, I arrive back at our barn to find our useless hotwalker leaning comfortably against the wall, hands in his pockets. When Jerry sees me , he remarks, "I guess I better get some bath water set." He moseys off to start filling pails.

I am livid. I have busted my ass running back to back races while Jerry has managed to do almost nothing. My stall is unmucked, the cooler is nowhere to be seen. Sandy's friends are still cooling out my two year olds we ran earlier. Sandy herself is already **picking stalls** for afternoon feeding. I still have to feed and do up all three fillies. With an effort, I hold my tongue.

I hurriedly slosh the water on Vera and scrape her off, still fuming. Jerry starts turning left but a few turns later, he is dumb enough to make some remark under his breath about how I am working. That's it! I lose my temper on the him. While I never raise my voice, I tell Jerry in no uncertain terms not to tell me how to do my job since he obviously cannot do his own. Jerry finishes the trip around the barn, throws the filly in her stall and quits. Hallelujah! I simply put the shank back on Vera and finish cooling her out myself. The two girls have finished cooling out their fillies. After grazing them for a few minutes, they quietly put them away. They offer to help but Sandy and I have everything under control.

Next morning, Mark is pretty upset with us. Instead of hiring a new hotwalker, he lets Sandy and I pick up the hotwalking. I am so pleased to be rid of the kid that it does not bother me in the least. With only six horses in the barn, there is no need for the hotwalker.

One of the two year olds develops a 'knee'. I cannot see the faint difference in the swelling of the knee, although many of the older grooms can easily pick it out. Mark decides to **tap** her. Tapping involves sticking a needle into a fetlock or knee to draw off the excess joint fluid (called synovia) that accumulates as the joint is damaged. The straw coloured fluid is squeezed out of the joint. When no more fluid seeps out, cortizone is injected into the joint. It is such a common procedure at the track that all our vets are competent at tapping.

I find this practice barbaric. Heaven forbid we give these animals time and rest to heal. Cortizone initially provides great relief but ultimately hastens the destruction of the joint. The wisdom of the track is that a horse can only be tapped three times before they break down. A sudden improvement in a horse's race form may be due to tapping.

When our vet pokes his needle into the filly's knee, the pressure is so intense that the synovia shoots clear across the stall. All of us exclaim aloud. It is a miracle this filly can gallop at all, never mind run. She must have been in a lot of pain.

Once the colts have all been cut and healed, I get few opportunities to ride at the track. I cast around for other chances to stay on horses. My friend Rhonda owns a fourteen year old stallion she calls The Old Man. She keeps him as a riding horse away from the track and now lets me ride him. Although he can be quite excitable around other horses, I love working with him.

After schooling him a bit, I begin taking Sandy with me and start teaching her to ride. She does so well in the limited confines of the round pen I soon think she is ready for a greater challenge. I take her outdoors for a lesson.

All is well for about ten minutes. Then my decision comes back to haunt me. The stallion deliberately drops Sandy, sending her sliding over his shoulder. As she falls underneath the horse, he steps on her chest and she passes out. The stallion gallops off towards the road. Oh, Lordy!

For a few seconds, I am not sure what to do. I must find and catch the horse before he hurts himself but I cannot leave Sandy slumped over in the grass. She starts moaning. I run to my car and drive it over to her. She regains just enough consciousness that I help her into my car.

At the end of the driveway, I encounter a couple of boys on bicycles. They point the way the stallion ran. I find him only a quarter of a mile down the road, running back and forth in front of a horse in a field. I catch The Old Man easily. By now, Sandy has fully come to. She slips into the driver's seat and coaxes the car down the road in first gear while I lead the horse. I put the horse away and drive us home.

It is not until that night I realize the damage the horse has done. Sandy lifts her pyjamas to show me the perfect imprint of a hoof on her chest. I am both horrified and mortified.

Worse, next morning, she is too sore to go to work and I have to confess to Mark how she got hurt. Now he is really annoyed with me. This is the first time Sandy has ever taken a day off. I am in the dog house.

Near the end of the Woodbine **meet**, Veracee has improved so much that she breaks her maiden for $10,000 – and is claimed. I cannot believe my ears when I learn we are planning to claim her back. It turns out she is a sentimental favourite of her owner. If you claim a horse and plan to run it within the next month, you must run it for 25% more than you claimed it for. Around the track this is known 'as being **in jail**'. The other option is to wait out the month. Horses sitting in stalls unraced do not earn any money so it is a tough choice.

Two weeks later, we claim her back for $12,500. By this time, we have moved down to Fort Erie. I have to ride the transfer van back with her. The filly has lost a noticeable amount of weight already. Obviously, the groom never catered to this filly the way I have been doing. Outside our barn, I stop to let her graze for a while. It is a lovely evening and I savour this quiet time with her.

The mood must be catching. Three fat geese stroll over to us and lay down on the grass, keeping us company. Geese? I have no idea where they have come from, nor do I see them again. They just seem so incongruous, lying there with me as the sun sets.

When Sandy and I relocate to Fort Erie, we move into tack rooms on the backstretch. Now we are required to do the night watering and we change Vera's water for the four times each day. Vera immediately puts on weight and begins to act like a racehorse. She flies around the end of the shedrow and puts in a few bucks for good measure.

Since she is still in jail, we must run her back for an even higher **tag**. There is no race for her at $16,000 so we put her in for twenty (thousand). Unbelievably, she finishes fourth, just a couple lengths shy of the winner. Pretty amazing for a horse that was soundly beaten for $6,250 just months before.

Just after we move to Fort Erie Mark decides to take a day off. He wants me to gallop Bo for him.

I am both excited and terrified. I have seen this stallion run off with Mark even in **draw lines**. The six year old horse is so strong that, even with a chain over his nose, he can yank me over to a bale of hay, grab the hay with his teeth and swing the bale clear across the aisle. Sandy has rubbed the horse for years and assures me she can help me.

It is pitch dark when Sandy leads me out to the little used training track on the other side of the road. She walks me to the centre of the track and turns me in the right direction.

"He'll come to me when I call him," she assures me.

Despite my nerves, I start off boldly. This horse has no mouth and he pulls unrelentingly. I use every tactic I have learned. I breathe steadily, brace my hands on his neck, straighten my back and use my whole weight. It is the longest mile. There is simply no letup.

In the dark, I cannot see Sandy. I am not even sure where I am. Fear claws at my throat. What if I can't hold this horse? Sandy hears me. When I am just about to pass her, she calls the stallion. He immediately softens underneath me. I guide him to the outside rail and he begins to pull up. Sandy grabs hold of him while he is still jogging. Within seconds, she has the shank on him and I can finally relax.

Every muscle in my body aches.

The next day, it is worse. I am so stiff and sore, I can barely move. This is only days after Sandy got dropped by The Old Man and she is equally sore. For the next couple of days, we move like a couple of little old ladies, shuffling bent over. It takes both of us to carry a pail of water. And Mark has absolutely no sympathy for us.

Chapter Twenty

THE SCAMP

Racing is all about prestige, winning races and making money. The cream of racehorses run in stake races (money added). In order to give the rest of the racehorses a place to compete, the majority of races are claiming races of differing amounts. Every horse in a claiming race is for sale for the amount of the claiming price. This allows trainers to quickly sort their horses out as to their value and hopefully make money with all of them. Horses that do not make money do not pay for their groom's wages and are eventually sold to be turned into riding horses.

Claiming horses is a part of life on the backstretch. Any time a stall becomes empty, either because a horse is sent home or is claimed, it needs to be filled quickly, otherwise, the stall will be reassigned to another trainer. There are a limited number of stalls on the track and many horses are kept on farms ready to fill any empty stalls.

Claiming races are the cause of much of the secrecy that is part of racing. If a horse is ready to win and outside people are aware of it, the horse may be claimed and the betting odds will be lower. If the horse is no good or breaking down, dumping the horse into a race may result in being able to unload a useless horse for a quick and lucrative sale. No trainer wants to tip his hand when he is making a claim.

Mark informs me that we are claiming a horse this afternoon in the third. I take a halter and shank and follow the horses to the paddock for the third race. I arrive at the paddock just as the horses are leaving it for the post parade.

Today there are a couple of claims filed for the same horse. I learn now that as the groom, I am supposed to be there to witness the draw. I get a scolding from the paddock judge for being late. I step into the foreign world that is the jocks' room. I pretend not to notice the skinny little men who wander around with only a towel wrapped around their lower bodies. I feel distinctly uncomfortable, as if I am in some forbidden territory. The judge assigns each of us a coloured ball. He places both in a tube and shakes one of them out. My ball is the winner. I am given a receipt. I escape to the paddock to await our new purchase.

The moment the starting gate opens, the horse is officially ours, even if he breaks down. Any prize money is retained by the former owner. I can hear the race being called, then there is a few minutes of quiet as the horses pull up and gallop back. The jockeys walk back through the paddock, carrying their saddles, with dirty goggles dangling about their neck.

Almost immediately, a small dark gelding appears, being led by a young girl who is bawling her eyes out. She leads him into the stall where I wait. With the horse facing the back, she pulls the bridle off as I slip the halter on. Sobbing, the girl hugs him one last time. Then I lead him home.

The young horse prances and drags me all the way to the barn. With the way he is acting, you would think he had won his race.

His name is Let's Just Scamper. He is well named. He is a lively little devil with absolutely no stall manners. When I duck under the webbing to enter his stall, Scamp blocks me. He will not move out of the way. When I push him away to make room for me, he bites me. He is unruly to groom and won't stand quietly to have his feet picked.

He is also a weaver. **Weaving** is a stable vice in which the horse stands at his door all day and sways back and forth on his front legs. It is probably due to boredom and being constricted in a stall all day. Once habituated, the horse will probably continue for life. Weaving wears out the knee joints, makes holes in the floor and takes a lot of energy out of the horse, leaving less for racing.

Next morning, the vet arrives to give the horse a jug. 'Jugging' is the practise of giving horses a couple of cups of vitamins, minerals and electrolytes intravenously to help them recover quickly after a race. It has become standard practice, as trainers will do pretty much anything to improve their chances of winning but it is not without controversy on the backstretch. A lot of the grooms believe it increases the chances of a horse bleeding after his race.

The bleeding issue is a bit of a mystery the vets have yet to unravel. No one yet knows why some horses develop nose bleeds after a race. Usually, it is just a trickle of blood, sometimes more of a torrent. Occasionally, the horse bleeds to death, referred to as 'bursting a lung'.

If there is just a tiny amount of blood, grooms try to wipe it off before the track officials notice it. If the horse goes to the test barn, this is harder to disguise. If the officials catch on the horse is a bleeder, it goes on a list, must be treated and the condition corrected somehow. This is a real pain for the trainer.

I don't very often take an active dislike to a horse but I make an exception for Scamp. Why his previous groom loves him so much is beyond me. I see the girl watching me about the backstretch and I suspect she may have sneaked into our barn to visit the horse. I have never seen a groom so attached to her charge.

I have already learned that no one will last long at the track unless you can put some emotional distance between you and your horses. Every spring, a new crop of two year olds arrives at the track. We nurse them through snotty noses and bucked shins, finally getting many of them to the races. Most will be gone in a year and a half, either because of a lack of talent, injuries or retired to the breeding farms. It is heart wrenching to lose your favourites but this is the nature of racing. To last at the track, you have to be resilient.

We run Scamp back only a couple of weeks later. Standing by the finish wire, I hear my number announced that my horse has been claimed.

As I lead him into the paddock after the race, I find the same little girl who had tearfully said goodbye is there to welcome him back. She is ecstatic as she buckles the halter on, oblivious of his attempts to nip her. Scamp jogs alongside of her as they leave the paddock in front of me. She is still petting him and whispering happy tidings to him. I never see the girl again.

Usually, when you lose a horse in a claim, there is a sadness of parting with an animal that you have formed a relationship with. Not this time. All I can think is, "Good riddance. You can have him."

Chapter Twenty-One

LAZY

In late summer, Mark asks me if I would be willing to work on the farm over the winter and break yearlings. If I agree, they will advertise the breaking and obtain a few more to break. It is a good opportunity for me, guaranteed work doing the riding I like. Of course, there are the rest of the farm chores but I am fine with that.

By the first of September, I start commuting to the farm. The farm is a long way away, fully a two hour drive each way. The first morning, I meet the other girl, Regina, also a Humber grad, who has been working on the farm since the spring. Regina makes me a coffee, assuring me there is no hurry to get working. After the long drive, I am grateful for the break.

The yearlings are coming in from the sales. The yearling sale at Woodbine runs the three days after Labour Day (first Monday in September). Most owners immediately ship them to a farm to be broken, usually a six week process. Then the youngsters get a couple of months off before starting into serious training in January.

Within a few days, the farm begins to fill up with young horses. We soon receive a dozen horses to break, doubling the number of horses on the farm. We feed, turn out and muck all twenty-four stalls in the morning before turning our attention to the yearlings in the afternoon. Regina proves to be an excellent ground person, following instructions to a 'T'. She is also decent at ground driving.

In that first week, we get three fillies in from the sales that have been together all along. We turn them out in the big pasture in the middle of the farm. However, when we go to catch them, not one of them gives the slightest glance our way, easily evading us. They have been in a stall for a full week without turn out and are busy eating. Thoroughbreds are usually easy to catch, having been accustomed to a routine of coming in since birth. Today, though, we are out of luck. When I leave at the end of the day, they are still out there. It will be sunset before Mike is able to catch them.

Mike is Mark's brother and the owner of the farm. When he bought the farm, it was being used as a riding stable. While it has been converted to a thoroughbred farm, Mike's daughter has a riding horse here and has a few jumps stored in the riding arena. The arena is small but functional.

I am pretty much in charge of the breaking. Since none of the babies are named, we quickly develop our own monikers for the young horses. Rat is one of the first babies I ride. Everything goes well as I walk and trot her. Then I try to stop her. When I pull on the reins, she flings her head straight up in the air and runs away with me.

I have absolutely no control. Since she has just been backed, her steering is non-existent. I gallop madly about the ring, sitting tight and praying that the filly will neither buck nor run into a wall. It is unheard of to be 'run off with' on a baby. They are usually so unfit that one spends the entire ride pushing them forward. After several laps of the arena, she runs out of steam and finally stops of her own accord. The next day is an encore performance. With that, we get the vet in and **float** (file) her teeth. We give her the weekend off and by Monday, she is fine to ride – and stop.

When I mention to Mark how Rat ran off with me, he comes out to ride with me and gives me some great pointers. He shows me how to sit back with extremely long reins and use mostly my stick and legs to control the babies.

Several weeks later, I have a scary incident on another filly. I have ridden this filly several days without incident and am not expecting any trouble now. Yet as Regina leads her out of the stall toward the arena, tacked and ready to ride, she explodes without warning in the aisle. It is as if someone has thrown a switch marked CRAZY. This is not a few mild crow hops but a full blown rodeo ride. She flings herself into the wall, eventually stopping. Nervously, we lead her out to the arena. I climb on carefully and have Regina lead me about to ensure my safety.

A few laps of the arena later, she goes nuts without warning again. She tosses me off on the very first buck and continues to pitch madly just inches from my body. I am knocked to the ground in one of those painful jarring landings that takes a few seconds from which to recover. Regina looks on, helpless and afraid that I will be jumped on and seriously injured but I'm able to roll away and get up. I am shaken but unhurt and I remount. Neither of us has any idea what sets this filly off and never will. I continue to ride her and we never have a repeat of this ugly surprise. We get the filly broke without any further incidents.

We later hear that she has a loopy streak in her even at the track. She has several repeated bucking sprees. I am not surprised. Horses' personalities quickly show up in their breaking process.

Eventually, I am getting on twelve babies, in addition to mucking, feeding, turning out and bringing in. This is on top of four hours of commuting daily. We still start our day with a coffee. By the end of the day, Regina and I are so tired we can hardly drag ourselves from the farm. We make a habit of hitting the coffee shop in town for a long relaxing break before heading home.

There are usually a few customers in the restaurant when we arrive. It doesn't take long before they clear out. We stink to high heaven. In order to pack all the manure on the wagon each morning, it is necessary to climb up on it and stomp it down. By the end of the day, we are pretty odoriferous. It is a game for us to see how fast the customers leave. Usually it is less than five minutes. I am surprised that we are not asked to leave instead. I nurse my coffee for half an hour before setting out for the long drive home.

Thankfully, I am driving against the traffic, which is pouring out of Toronto.

Some of our racehorses begin shipping home. They are full of surprises for me. On the first day home, I expect TriFran to explode when I turn him loose in the arena since he was so hyper at the track. Yet, when the two year old is let loose in the arena, he simply wanders around aimlessly.

Dum-dum, on the other hand, stays true to form. I had been told the gelding jumped the five foot gate from a standstill the previous year. When I turn him out by himself the first day, I don't expect him to jump out of his paddock immediately. Just twenty minutes later, Mark walks into the barn and says to me, disgustedly, "Go get Herman. He's running up and down between the paddocks already." I catch the big gelding and he is relegated to arena turnout only.

Dum-dum hasn't shown any talent for racing but he can obviously jump. He is put up for sale as a jumping prospect. A week later, I am asked to show the horse to the prospective buyer. I saddle him up and set a **cavalletti** adjacent to the wall in the arena. Now, Dum-dum has never schooled over so much as a pole on the ground so I take my whip and basically beat him over the little jump, all of sixteen inches high. Faced with this tiny object and this mad woman on his back, the gelding makes a huge leap over it.

"I want to see it again," the man says.

So I circle back around and push him over the cavalletti again. After the twentieth time, I refuse to jump it once more. "Look, we told you he has never been schooled over a jump so I am not going to keep beating him over this. He has jumped out of the paddock at least twice so we know he can jump. You will have to make your decision on what you have seen."

I put the horse away. I've done my job. He decides to buy the horse. A week later, he arrives to pick the horse up. To my horror, he is pulling a converted U Haul trailer just six foot high, not a proper horse trailer. I have seen too many claustrophobic racehorses flip out in two horse trailers. The new owner drops the ramp and I see there is already a full haynet hung in back. We use old wraps to ship Herman, knowing we will never see those again. Tentatively, I lead him up to the ramp. The huge horse willingly follows me into the tiny trailer and tears into the hay as I tie him up. There is so little room that the haynet is squeezed onto my left cheek and the horse is pressed onto my right cheek. He cannot even lift his head up. His withers barely fit. We shoehorn him in and seal the trailer tight. There is no light, no ventilation. I cross my fingers. If he blows a gasket now and tears the trailer apart, there is no room to get in there and untie him.

Bless his heart. With the hay in front of him, Dum-dum settles down for the long ride. He is off to Montreal, four hundred miles away! He is one of the few racehorses I know that would even load in such a small space, let alone travel ten hours successfully in that sardine can. Eventually, I hear the Herman has become a sucessful jumper but I never learn what his show name is.

I am still getting on a dozen youngsters. They are incredibly nice horses, well bred, athletic and good looking. One filly in particular impresses me. From the very first ride, she frames up and feels like a horse that has been going for months. It is so easy for this girl to carry me. However, I am learning the down side of her athleticism. She has a tendency to cut her corners. Once, I try giving her a tap on the shoulder to encourage her to go deeper into the corner. Instead, she instantly wheels one hundred and eighty degrees and I plow head first straight into the boards. I cannot believe a horse can move that quick.

By the end of the day, I am so exhausted I can barely drive home. Our morning ritual of coffee before work has stretched longer and longer, with a second coffee break when we finish mucking. For some reason, Mike is convinced I am not pulling my weight. I tell him that riding twelve babies is like riding twenty four regular horses. I am mucking a dozen stalls, feeding and turning horses out as well.

Regina suggests she might be able to get on some of the horses. I am quite sure that her skills are minimal but there are a few quiet ones. I check with Mark first and let Regina start riding my three quietest horses. That is a full hour less of riding for me. With

Regina riding, we can train some of these babies together. One of the colts proves to be a real chicken who will wheel out of a tight spot. By now, I have developed a Crazy Glue seat and am ready for those shenanigans. With some practice, the colt learns to handle the tight spots. He will need that skill for racing.

It never occurs to me that Regina might be jealous of me. She had been a model employee before I started work but has been quietly slacking off. She encourages me to take long coffee breaks and I am gullible and tired enough to go along with her. Mike, though, thinks I am the cause of this. Before I started work on the farm, Regina looked after a dozen horses by herself.

One day, Mike and I get into an argument. He calls me lazy, which infuriates me. I defend myself, having worked harder here than anywhere else I have ever worked. Nevertheless, just before Christmas, Mike fires me. He decides that I am a bad influence on Regina, that she accomplished much more before I arrived at the farm. Regina, smirking, waves good bye to me as I drive away for the last time.

Chapter Twenty-Two

MAKING ENDS MEET

"God damn flies!" Doug, my new trainer, sits hunched between us on the fat gray pony, waving away imaginary bugs with his hand in the frosty air.

Despite the raw February weather, Brydon and I smile back at Doug's attempt to cheer us up. "Cold, eh?" Brydon and I nod in reply. It is so bitter that my nostrils stick together when I suck air in and the vapour from the horses' breaths frosts their whiskers.

I am glad of this chance to get on horses, even if it means galloping this early in February. My weight is still working against me. Many trainers will only hire girls that weigh one hundred and twenty pounds or less. Packing on the thirty pounds when I broke up with The Boyfriend is hurting my chances of galloping at the track. Still, at this time of year, there is a shortage of exercise people willing to brave the cold. Doug overlooks my weight and lets me exercise the horses.

I am never sure what to make of Doug. He likes to keep us guessing about his intentions, in that if he shows up late and we start to get the first set ready, he will send out another pair of horses. I wonder also about his past. He once trained horses for one of the most prestigious outfits on the grounds. Now he has a smaller stable he runs. Usually a top trainer is in demand and is courted by the more wealthy owners. What happened that he has come down in status?

Doug certainly knows his horses. He says little and thinks much. Every day, he rides out with each set between Brydon and me on a chubby little horse. Once we are at the track, he gives us our instructions. After Brydon and I pull up, we all ride home exchanging pleasantries but with no comment from Doug on how our horses did. I am never sure how he feels about my skills.

I am not sure why Brydon is out here galloping with us either. He is a young jockey and most of the jockeys go south for the winter. He must be hurting for money to work here in the cold; but on the other hand he is easy to work with. We have both been around the track for a while and have developed the off beat humour characteristic of track veterans. It is comfortable to have such an easy work mate. There isn't anything else that is easy for me right now.

It is brutal galloping in this weather. The wind cuts through our cheap knit gloves, the thin ones which work well for galloping. The reins slide easily through thick, warm gloves, allowing horses to tear off uncontrollably. The protective goggles we must wear fog up but without the goggles, the cold temperatures and wind cause tears to stream from my eyes. Galloping in winter becomes a test of endurance. Still, I am grateful to have this job. I also have another which is helping me financially.

It was only days before Christmas when I was fired from the Mike's farm. Whenever I have been out of a job before this, I would just return home to my parents to sponge off them. Now, with my name on the lease of my apartment, I have to pay my half of the rent regardless of my financial situation. Late December is a bad time of the year to be out of work so I hit the pavement and start filling out job applications. I will do what is necessary to pay the bills.

I get lucky. I stand at the counter of the Ponderosa Restaurant, smiling hopefully every time the manager looks up from the grill. When I tell him I am looking for work, he asks if I would be willing to commute an extra half hour, I'm so desperate I give him a resounding 'yes'. He hires me on the spot and as far as I'm concerned, after the two hour commute I just finished, an extra half an hour drive will be a breeze. I breathe a sigh of relief to have full time work. As it turns out, the hours suit me perfectly. I am trained as a cook and work from twelve to eight. My day goes by in a whirl.

I have never punched a time clock before. I love that I can actually leave at the end of my day instead of being tied to any barn at all hours. I also find it a relief that I no longer have to fend off the men with the groping hands and their sordid attempts to get me into bed under any pretext.

I settle comfortably in my new job. The time management skills I picked up while working at the track serve me well. When the district manager comes in to inspect the store, he watches me closely, then asks me how long I have been working on the grill.

"Let me see. One, two, three, no, four days."

He can hardly believe it. He thought I had been doing the job for weeks. Within a couple of months, I earn employee of the month and start training the new staff.

This job affords me more free time. While breaking yearlings in the fall, I had neither had the time nor the energy for recreational riding. Now I resume riding Rhonda's stallion in the mornings before work. I have a secret desire to take the horse to some shows when he and I are sufficiently schooled. Rhonda has moved the Old Man to a barn with an indoor arena. This allows me the opportunity to continue riding him through the winter. Being an ex-racehorse, the horse still gets wound up easily and he has only the most rudimentary training. I decide not to canter him for the time being and work on his basic schooling. This also affords me a chance to work on the weaker elements of my own riding.

When the track opens back up in February, I apply for a galloping position. Exercise riders only work the morning hours, unlike grooms who must return for afternoon feeding. Many exercise riders hold down a second job, supplementing their income. Doug, a trainer who is unknown to me, needs another exercise rider and takes me on. I can work both jobs, provided I am away from the track by 11:00. However, working two jobs does not allow me the time for Rhonda's stallion and I quit riding him for the moment.

My new barn at Woodbine is beside the sand ring, an unsupervised eighth mile track favoured by the pony boys and those with youngsters not ready for the bigger tracks. Doug sometimes sends us out there on two year olds. Mostly, we walk the length of Woodbine on either side of Doug and gallop on the training track while Dave watches from his pony. The walk to and fro is no picnic. On one particularly windy morning, Brydon confessed he was crying behind his goggles from the pain of the chilblains.

One frosty morning as I am riding a two year old filly around the sand ring, she bucks me three feet clear of the saddle. Only the reins, which direct me back into the saddle, keeps me on. When I get back to where Doug is watching, I remark, "Well, I fulfilled my life's ambition. I always wanted to be a cowgirl." Doug's only comment is a smile.

Another morning, Doug sends me out on the pony to the sand ring to pony a little black mare that just came in. I'm excited to try ponying and know that Doug is watching from the barn; I've been hoping for an opportunity like this. I jump on the pony and ride him down to the end of the barn. I wait outside until the groom gives me the filly in my right hand. I'm steering the pony with my left.

I walk out to the oval and start off on an easy canter. The pony slowly picks up speed for one and a half turns. Then I pull back on the reins to steady him. Instead, the pony ignores my effort and continues to pick up speed until he is truly running away. With one hand engaged with the filly, I cannot get the pony back under control. We fly around the rest of the four laps before I am finally able to get pulled up. I can sense the disapproval of the other riders around me.

"How did you make out? Doug asks innocently, as if he had not seen my fiasco.

"I got run off with." I am feeling about 2 inches high.

Doug explains to me that my mistake was to take a steady hold on the reins. Instead, I should have used strong pulls and releases.

Now, however, Doug is concentrating on the filly, who is beginning to **tie up**. I feel horribly at fault. The filly is covered with a cooler, not walked and the vet is called. Doug is not worried. "She has a history of tying up. Tomorrow we will walk her and after that we will give her some Milk of Magnesia before she ponys. To my everlasting relief, Doug lets me try ponying again. Now armed with the knowledge of how to handle his rascal of a pony, I manage to keep the little 'bugger' under control and, while the filly is feisty, all goes according to plan. Ponying becomes part of my duties.

Doug now hires a small young guy straight off the street to hot walk. Because of his youth, we dub him 'the Kid'. He also wants to learn to gallop.

Doug brings a new two year old in from the farm. This horse has never had its mane pulled. With his long mane and forelock falling over his eyes, we promptly name the horse 'The Mustang'. Brydon concocts a lie and tells the Kid that Doug is going to put him on the Mustang. I see the Kid's eyes widen with fear. Brydon goes on to embellish the lie by telling him how dangerous this horse is. In collaboration, I faithfully back up the story. Almost on cue, a few minutes later Doug does tell us to get the new colt ready and put the Kid on the Mustang. The fear level in the Kid ratchets up. Both Brydon and I are aware that Doug would never have put an inexperienced rider on this horse unless he is dead quiet. However, we are not about to give up the game just yet.

By the time the colt is tacked up and I lead him out into the aisle, the Kid is (pardon my French) 'shitting bricks'. I boost the young guy up in the saddle and lead him around the shedrow while he ties on. His hands are shaking as he knots the reins, tightens the girth and adjusts the stirrups. He begs me not to turn him loose just yet. I take another turn of the shedrow before letting him go on his own. Of course, the Mustang is as gentle as a lamb. Brydon and I are howling with laughter. Doug has not cottoned on to our private joke or he would skin us alive for scaring the Kid so badly.

For two months, I work both at the track and at the restaurant. The galloping is seven days a week. On weekends, I rest up for my punishing schedule. I am pulling eighty hour weeks fueled on coffee to keep going. I train the waitresses at the restaurant to never let my coffee cup run empty. At night, I collapse into bed with exhaustion. At least my money worries are banished for the time being.

As the weather warms, the population of the backstretch grows like a rabbit colony. Help is now plentiful. All this time, I have been helping out around the barn, galloping, ponying, walking hots and mucking the odd stall. Now Doug hires another exercise rider. He asks me to groom instead. My weight is against me and I know I will not find another job galloping this season.

Doug insists that either I give up the restaurant and groom full time for him or quit altogether. I choose the restaurant because I am learning new skills on that job.

For a month, I sleep ten to twelve hours a day to balance my sleep deprivation. Then I resume riding The Old Man, Rhonda's stallion. Although I ride four or five mornings a week, I miss being outside all day long in the glorious days of summer. At the restaurant, I earn a promotion into management ranks. Even though the management can be very arbitrary, the training is extensive but my work schedule becomes more erratic. I console myself that this is a good opportunity to learn management skills which may further my career.

On my days off, I still visit the track and hang out with my friends, lending a hand if they need one. My girlfriend Sandy is now rubbing Tri Fran, the colt I love so much. He begins to break down and Mark drops him into the claiming ranks to lose him. The colt is claimed by another barn. The next time Tri Fran runs, Sandy happens to have a horse in the same race. I am helping her out that day. When she returns to the barn after the race, she is swearing a blue streak.

"What happened?" I ask.

"It's Frannie. He broke a leg and they put him down"

Later, I walk to the back of the track and find the horse's body lying behind a barn, already covered in flies. I feel sick, disgusted and angry. I truly loved this colt and wish I had been able to buy him as a riding horse, but the $15,000 price tag was way beyond me. Now he is dead, no good to anybody. It still pains me to think of the utter uselessness of seeing good horses destroyed. Why couldn't they just given him to me?

As a junior manager in the Ponderosa Restaurant, it becomes apparent that the company is struggling for its survival. The head office is panicking and making poor choices that only hastens its demise. I am yanked out of the store where I had such a good relationship with my boss and moved to three different restaurants in a matter of months. My performance is good under one manager. I do not measure up in the second restaurant and I manage comfortably in the third.

This brings home to me the importance of working for people I like and respect. Under encouraging people I shine but without support, I just crumble. I am grateful for the management skills I am learning but I chafe at being inside all day, never even knowing what the weather is doing. I miss the track, the horses and the people. I realize that I am just biding my time. I need to go where my heart is calling me.

Chapter Twenty-Three

DOING WHAT IT TAKES

I wake up on my twenty fifth birthday crying.

I haven't the foggiest notion why I am so unhappy. I do not fear getting old. As an adult, I love being able to pursue any goal and make it happen. I am living my dream, working with horses, have my own comfortable apartment and am surrounded by friends. What the hell is the matter with me?

When I take a closer look at my life, I see I am working like a dog for very little money. I have no marriage prospects. My vague notion of working with horses has resulted in spending most of my time on the end of a pitchfork. I have very little control over my schedule.

I may not know what I want but it sure isn't this. The trouble is, I'm not sure now what I really want.

I am still riding The Old Man and the horse is going beautifully. I continue pushing myself to further my riding skills by riding without stirrups for twenty minutes every day, trying to achieve a decent sitting trot. By experimenting with my training, the horse is definitely exhibiting a much more polished performance. I wish the same could be said for me and that I could afford riding lessons. Show riding is a very different skill than conditioning racehorses.

Being a stallion, the Old Man is occasionally bred. Two weeks after servicing a mare, Rhonda joins me for a **hack**. She has a young filly in the same barn as her stallion that she is breaking. I tell her to be sure and bring a stick.

We ride no more than half the length of a field when we come to a small ditch. The Old Man easily traverses it but the filly wants no part of it. She plants her feet and refuses to cross. Now isn't a good time to learn Rhonda hasn't brought a stick.

We ride back to the barn where I insist on trading horses. I tie the Old Man in his stall while Rhonda waits in the arena. She gives me a leg up, then brings the stallion into the arena. He still isn't standing reliably to mount and as she goes to jump on him, he leaps away from her.

Instantly, the stallion realizes he is loose. The amorous thoroughbred makes a beeline for the filly I am riding, intent on breeding her. I smack the filly hard, sending her into a gallop. The stallion gallops alongside us, his nose glued to her flank, smelling her scent and excitedly calling.

If the filly stops, he will mount her. I either must bail first or be injured. This has just become dangerous. Using the infamous stick, I try to beat him off but he is not to be discouraged. Fortunately, the barn manager happens to be walking through the barn and sees the predicament I am in. He runs into the arena and manages to grab the reins on the stallion as we race by. As quickly as the event unfolded, it is over. The barn manager gives Rhonda a leg up. As soon as she is in the saddle, the stallion settles down and behaves elegantly. It is as if the incident never happened.

We go for our hack. The Old Man has one ear forward but the other ear is still latched onto the filly. When he begins yelling, a quick smack on his neck reminds him to mind his manners. It is a fun ride.

Despite working full time at the restaurant, the track calls me back. When February rolls around again, I drive through the gates like the track veteran I am, pausing only long enough at the security gate to give my last name. I make my way down to Doug's barn and he promptly hires me back.

Brydon is gone and the new exercise boy, Jimmy, has less experience than I have. He is young enough to be impulsive and an egotist, not the safest attitude for galloping racehorses.

I am fully settled into the routine of racing. The seasons roll one over one another with one year much like the one before. Spring brings the optimism with the young horses, racing is in full swing by summer, in autumn the barns begin to shrink and for the winter the horses move to the farms. Time moves in circles at the track with the same jobs performed by the same people. It is only when I run into old school chums that I realize my former classmates have moved on, married and are having children. My life is at a stand still.

Again I am working two jobs, logging up to eighty hour per week. My new restaurant manager kindly assigns me closing shifts only, allowing me to gallop at Woodbine. By the time I close up on a Saturday night and drive home, I've only had four hours sleep. Once more, I train my staff to never let my coffee cup run empty. The girls obediently refill it every time they walk by. I must be drinking twenty cups of coffee per day.

One Saturday night, some customers linger late. Only when the last of my staff have punched out do I get down to finishing the restaurant records. I am tired enough to keep making mistakes. One night, it is three A.M. before I lock the door and drive my car out onto the street.

As I pull out, a police car passes me. I just sigh when he flashes his lights and pulls me over. Suspicious as to why I am leaving the restaurant so late, he asks for identification. When I reach for my purse, it isn't on the seat. Damn! I left it sitting on the desk in the office in the restaurant.

I have to talk my way out of trouble. The cop believes me and lets me go. Now I'll only have one and a half hours sleep this night. Great.

I cannot continue to work these crazy hours. I am always exhausted and using coffee to fuel myself. Even when I wake up, I am still so tired every morning.

Doug still frustrates me, making me late for work by never letting me leave in a timely fashion. He always drags the morning chores out and finds more for me to do at the last minute. I am not quite settled in either world. I feel out of touch in my own barn.

One morning I show up to find an empty stall instead of the big bay two year old that I gallop. I learn there was an accident in her race yesterday. One of the horses went down in the race, sending her jockey onto the track in the middle of this melee. A second filly fell over her, kicking the jock's helmet off. Then our filly, who was about 1200 lbs, fell on top of the downed jock.

The jockey lies in a coma for six weeks. He has a shattered shoulder blade, fractured skull, broken ribs, etc. He hovers at death's door but pulls through. The horses weren't so lucky. All three were destroyed.

It is a fact of life that sooner or later, every jockey will be involved in a catastrophic spill. It is a case of when, not if. The Jockey Club does all it can to make sure the horses are sound to race and the track is reasonably safe. But accidents happen.

In April, Doug hires another exercise rider and offers me a choice. Either groom full time for him or be laid off. If I stay, I will still get to pony.

I debate about the decision. While I am glad of the experience of managing a restaurant, I can plainly see the chain is in dire straits. There is little future there. With my weight still against me, galloping for another outfit is not a viable option. I make a different decision than last year and go back to grooming.

I'm frustrated but I don't see any other choice.

While working both jobs, I had no time to ride The Old Man. Over the last couple of years, riding on my own, I had improved the horse considerably. I hope to find some way to take him to a dressage show.

Once I quit the restaurant, I have time again to keep my riding skills sharp by schooling Rhonda's stallion. However, it is as if I have never worked with him. Instead of the well broke horse of a few months ago, he is uncontrollable in the canter and unable to concentrate on my aids. Two years' work is down the drain. I am so discouraged that I haven't the heart to start all over again from scratch. Still, I have learned a lot about training riding horses, improving my flat riding in the process. The horse also provided a good opportunity to practice teaching riding. I realize that the horse was not wasted time.

Now that I am at the track all day, I get to know the owner of our barn. Nick is a self-made millionaire who built his fortune by literally mixing mortar and stacking bricks. This hard working-man is keenly interested in his stable and believes in treating his help well.

When we win a race, he immediately drives back to the barn and hands out crisp pink fifty dollar bills to each person in the crew; hotwalkers and other grooms included. When I return to work the morning after a win, I too, am handed a fresh pink bill. Later, Nick will check with me to ensure I received the bonus. After years of being stiffed for my stake money, this generosity is well appreciated.

The other groom in our outfit is Pee Wee, a creepy character who stands under five feet tall and spends his afternoons frequenting the strip joints. I feel sorry for the strippers. Personally, I wouldn't want him anywhere near me.

Pee Wee and I alternate feeding in the afternoons if we are not racing. On my first afternoon off, I never give it a thought and hang my haynets up high as usual. Next morning Pee Wee gives me an earful. He couldn't reach my haynets to take them down. I have visions of the little gnome leaping madly at the string and failing. When he turns his back to me, I permit myself to smile.

Doug now brings his teenage son with him to hotwalk for us. The kid is somewhat mentally challenged but hotwalking is well within his capacity to handle. Doug fiercely protects this boy, making sure we never take advantage with thoughtless pranks. This is a totally different side of Doug, one I never suspected.

One of the horses I am given to rub is a four year old mare named Raphael. She is a real handful. Because she hates to be groomed, I only lay her hair flat with a soft brush after bathing. Many racing thoroughbreds develop ticklish skin but most will at least stand to be bathed. Not this mare. She kicks so freely that I am forced to hold up a front leg while I bath her, trying to keep her hind legs on the ground. At feed time, she runs around her stall excitedly while kicking out. I just wait until she is facing the entrance before ducking under the webbing to hang her tub. I absolutely adore her.

She is one of the horses I regularly pony. The pony is so slow that I develop saddle sores on my butt (this is the same pony that ran off on me the year before). I invest in some Elastoplast bandage and paste strips of it on my sore bottom. I hope I never have to go to the hospital with these strips on; a nurse might think it some weird fetish of mine!

I often pony Raphael on the training track in the latter part of the morning. She is quite challenging, as she leaps into the air and throws herself against the pony as I gallop. When I pass the gate where horses are schooling, she is even more difficult.

The guys in the gate are extremely careful about not letting horses out of the gate until there is a clearing of the traffic. This avoids a possible collision or causing other racehorses to run off alongside the working horses. Ponies dont't get this consideration. Just as I swing by the gap where the gate is located, I can bet one or two horses will be breaking out of the gate past me. And sure enough, it happens as if one cue, causing Raphael to go nuts. My pony also tries to grab the bit and run off. I learn to handle both horses at once, no mean trick. It is a skill I will be glad of in my later life.

The mare is difficult in other ways. It is no fun to paddock any horse on a hot summer day. With a fiery horse like Raphael who leans all over you, the salt in the sweat bites into your arm, causing abrasions. After the race, the dirt from the track is mixed with the sweat, burning even more. The first time I run Raphael, she sweats up so badly that by the time I hand her off to the pony boy in post parade, the water is pouring off her. I realize she has 'left her race in the paddock.' Sure enough, she runs poorly.

This is a puzzle to be solved. I need this mare to get to her races as relaxed as possible. There are a number of tricks I can use.

I give Raphael a quick bath with some rubbing alcohol in the water before I go over for her race. I leave her wet to help her stay cooler. I arrive at the tunnel at the last possible moment to not to upset her by the other horses leaping about.

My last trick is to walk ver-ry slowly to the paddock. When I arrive at the paddock, I pretend to be surprised by my lateness ("I left the same time everybody else did") and get there just in time to be saddled. It works. She doesn't sweat up and her race performance improves considerably.

There are still reminders of my past around. As I am mucking stalls one morning, I glance up to see J.B. riding a horse down the shedrow. I have not seen the old man since he fired me three years ago. He is unsure of whether or not to speak but I say 'hi' and he returns my greeting a little uneasily. I harbour no grudge toward him.

A couple of days later, I see J.B. ride by, flushed and sweating heavily. He says good morning, despite his laboured breathing. Two minutes later, one of the grooms from the far side of the barn shouts, "Some old man just died." I instinctively know it is J.B, that he was having a heart attack even as he spoke to me. I rush over to the other side of the barn. In on open stall, l find J.B. face down in the stall, the saddle still in his hands. A groom holds the horse in the back of the stall while a couple of us try to do CPR. It is too late; he is already dead. The paramedics arrive shortly and assure me nothing could have been done. I return to my barn somewhat shaken. It is not easy to see death so up close and personal.

One morning, everything in our barn changes. Nick, the owner shows up early in his huge Cadillac. Doug has not arrived. Nick parks his car in front of the barn and, one by one, ushers the help into his 'office'. When I slide into the seat, he apologizes for the mess of papers, then gets right to the point.

"I am letting Doug go. I found he has been stealing from me. If I let him go, will you stay with the outfit?"

Although I like Doug, I have no personal loyalty to him. It is Nick who pays me. I have no hesitation in replying, "Sure."

Just like that Doug is gone. I never see him again. So often at the track, people just move from one outfit to another but I never hear of Doug.

Nick approaches Louie, the trainer of just one horse stabled next to us and asks him to take over as our trainer. Louie reluctantly agrees.

Everything Louie does is reluctantly. The poor man is in way over his head. Louie has no idea how to handle staff or organize the training. Each morning, we have to feed our new trainer ideas on what to do with the horses.

But this does give me an opportunity to experiment with my horses. I decide that Raphael needs to stand in ice every morning after training. I find an old muck basket and fill it with ice and water. As I am guiding her into the tub for the first time, Pee Wee is hovering over me. "Oh, she'll never stand in a tub," he predicts ominously. The next morning, the mare practically climbs in herself.

By early August, I have figured out how to keep Raphael sound and happy. Now that she is ready to race, she is eligible in three possible races, all in the same weekend. There is a minor stake race, an allowance and a non-winners of two. Nick opts for the stake race.

Raphael needs an easy work as a **tightener**. Nick goes looking for a jockey but the only rider he can find is an inexperienced apprentice. We are based at Woodbine while the jockeys have moved to Fort Erie for the A meet. As I have to lead the mare out to the track, I am able to watch the work.

It is a disaster. She runs away with her rider. In that first quarter mile, I have never seen a horse run that fast. The kid does manage to get her to slow down in the next eighth mile and the work sheet showed a quick three eighths in thirty five and change. The kid is hugely apologetic. But the damage is already done. Our race is left on the training track. Not only that but we hook a tough field. The best race mare on the track is entered against us. Too late Nick learns what a poor decision he has made. The good mare wins the race. We run up the track.

By the end of August, as per usual, the yearling barns are filling up with youngsters for the sale, held after Labour Day. Rhonda comes looking for me. Am I interested in making a little extra money? She knows a person who needs some help. Since I am perpetually broke, the idea of making some easy cash appeals to me. Rhonda takes me over to introduce me to the barn manager.

The manager thinks I look vaguely familiar. I, on the other hand, know exactly who he is. I pray his memory does not improve.

She runs away with her rider. In that first quarter mile, I have never seen a horse run that fast.

In my first job nine years ago, I was hired to rub yearlings for the sale. At sixteen, I was uncoordinated, socially inept and a general pain in the ass for those I worked with. I never knew when to shut up and was by far the worst worker on the crew.

All of the staff in the yearling division worked together mucking the barns. Bryan Rickey was the senior groom. He was foulmouthed, impatient and had big plans for his own future. I got under his skin from day one and things never improved. My slowness, my non-stop talking and my inability to muck a stall cleanly irritated him. Within a few days, Bryan-Rickey dubbed me Speedy and never used my real name again. That is, if he called me anything at all. Mostly he avoided me like poison. The more I was aware of his dislike of me, the more I tried hard to please him. It never worked. I was just more gawky, more guilt ridden than ever.

The wagon we mucked into had tall sides on it. As it filled up, we had to pile the manure higher toward the tractor end. One day as I threw the manure onto the wagon, half the manure fell off of my fork. The rest sailed over the side of the wagon to land on Bryan-Rickey's bald head. He froze in mid-stride. He was so mad he was speechless. I was terrified of what he might do and stuttered my apologies. I was also afraid I might just give in to this wild desire to laugh out loud. If I was 'persona non grata' before, I was damn near invisible to him after. I just fell off his radar unless it was to make some snide remark. Now, years later, the memory is as fresh in my mind as the day it happened. I am not about to inform Bryan-Rickey that I am the former 'Speedy'.

Bryan-Rickey is now managing a small farm and has four yearlings to sell here. The dark bay colts are as alike as peas in a pod. They are tall with particularly long back legs and few manners. He wants each of them hand walked for about 20 minutes while he mucks their stalls.

Colts are apt to rear or nip any time one quits paying attention. I have to stay alert the whole time I am walking them. Despite my experience, one of the colts manages to cow kick me in the stomach, even though I was walking him correctly up by my shoulder. A few people rush over to see if I am hurt but fortunately, the colt could only manage a glancing blow.

For three days, I hand walk the colts while Bryan-Rickey mucks the stalls and fills the water pails. At the conclusion of the third day, Bryan-Rickey pays me off and thanks me for the good work I've done. He never does make the connection back to 'Speedy', to my everlasting relief.

By the end of the summer, my barn claims a little bay horse named Stevie. He is sour, training poorly and running badly. I suggest we try ponying him.

By the third day of this, he pricks up his ears and dances up to the pony. He begins to act up as I pony him, bucking and kicking out. It is always good to see the horses so full of themselves. We run him a week later but he **comes up short**. Looking back, I see it has been three weeks since he worked. I point this out to Louis and we start to work him weekly besides ponying him. It works. He wins but I have already left.

Louie is just not working out as a trainer. Nick lets him go (I am sure Louie is relieved to be out of that job) and decides to take on the training himself. If Louie knows little about training, what does Nick know? I do not want to work for somebody who knows so little. Besides, summer is already winding down and I want to get riding work on a farm.

It has been a frustrating year. When I do ride, I ride well. I can improve horses' performances. It feels that for every step forward that I make, I take another step back. I am a long way from figuring out the solution for my tears of frustration.

Chapter Twenty-Four

GETTING MY BREAK

"I'm not coming this morning. I've quit".

I cannot believe what I have just heard Sally say. I am supposed to pick this girl up, as I have for the past four weekends and bring her with me to the farm where we are currently working for the winter. This morning, the two full time employees are heading off to a seminar in Toronto leaving me a chance to gallop some of the horses in front of Red. He is the prestigious trainer for this farm. Sally knows how badly I want to ride for Red, hoping the trainer will hire me for the track. Now I will have to do all the chores alone, mucking the stalls, feeding and turning out the other horses. Red, who is bringing his regular exercise rider Jackie with him, will simply put Jackie on all the race horses and I will not get the chance I so desperately want at galloping. Quitting now is pure spite on Sally's part.

I have decided that come hell or high water, carrying extra weight or not, I am determined to get back to galloping even if it seems to be an impossible goal. I try every trick I know to motivate myself. I recite positive affirmations, put a photo of me riding a racehorse beside the bed and visualize myself galloping. Now, with the goal in sight, I am threatened to be derailed by a spiteful brat.

Even getting to this point has been a circuitous route. Five months earlier, I searched for a farm job where I could advance my skillls, especially my galloping. I deliberately picked Wilbur's farm to work at since it specializes in **lay ups** and breaking, hoping to get more riding. The farm has a fairly large crew, evenly divided between women and men. My first morning there, I watch with disbelief as the girls ask the guys to move the tractor each time they finished mucking a few stalls. Having been part of farm crews for years, I know the last person finishing mucking their stall is supposed to move the tractor ahead. Yet not a single other girl on the farm drives the tractor. Because of this, the guys lord it over the girls about how useless women are.

I don't say a thing. For three days, I watch for my opportunity. Then, when it is my turn to move the tractor, I jump on the seat and simply drive it ahead. The silence is deafening. At least I do not have to endure the taunts of the guys.

Despite his reputation, Wilbur is a wheeler-dealer and an arrogant, self-made man who needs to be worshiped by his crew. He deliberately hires young staff who he can mold to his ways and in return, they will admire him. Only Manly and I, who are from the track, do not subscribe to the adulation of our boss.

One day, Manly and I take a couple of older horses out to ride around the property. Manly warns me to be aware of the filly I am riding as she is easily spooked. I am enjoying this break in the open air, away from the boss' watchful eye. When a low flying helicopter comes by, I forget Manly's admonishment and wave at the operator. In the blink of an eye, I find myself lying on the ground, staring at the sky. So much for improving my riding. I can't believe this filly can duck sideways so fast. Bless Manly's heart, he never tattles on me. I rarely get to ride and this embarrassing incident might keep me from riding altogether. Besides which, the boss demands a case of beer from anyone who falls off.

The horse owners at this farm are all middle class people, with one horse or two at most and are woefully uninformed about their animals. The farm does an adequate job of the breaking and layups, successfully hiding any issues from the owners. When I make the mistake of honestly telling an owner about his horse, I am yanked into the office and given a severe dressing down. Other than this, Wilbur just ignores me as much as possible.

Wilbur's personality is like sandpaper to me. When one girl turns a horse out exactly where Wilbur tells her to, she then has to endure his abuse that she has made a mistake. This girl quickly backs down and owns 'her' mistake. I am not that diplomatic and my ego is too big. Again, I am doing mostly ground work, mucking, bandaging and moving horses in and out. While there is quite a bit of riding to be done, I only get a small part of it. My days here are numbered.

One of the senior grooms owns a horse in partnership with Wilbur. The horse is laid up with bad ankles. After a couple of weeks with no improvement, I ask the groom for a free hand to treat this horse. When given the okay, I set to work, using the methods Sandy taught me. I alternate poulticing, sweating and alcohol braces. The ankles start to improve. I even leave instructions with staff on my day off. Although the horse improves significantly, Wilbur never acknowledges my contribution to his horse.

Yet Wilbur seems to trust my skill level. When I knock on the door of his house one afternoon to fetch him for a horse that is beginning to colic, he does not answer the door. Instead, he yells through the closed door for me to give the medication intravenously.

I never have done an intravenous injection. At the track, only vets are allowed to handle needles. However, I have watched dozens of time and know the protocol and the precautions. I carefully block the jugular vein with my thumb, insert the needle as vertically as possible, draw a little blood into the syringe and slowly depress the liquid. I am quite pleased with myself. This, however, makes no difference to how Wilbur feels about me.

I get an occasional opportunity to help swim the lay ups in the pool the farm has. Swimming is an excellent way to condition an older horse with leg injuries. It provides a good cardiovascular workout without putting stress on the legs. I find this activity fascinating. The pool is shaped like a thermometer with a bulge at each end to turn the horses. The horse is led down the ramp, then the bridge is dropped behind the horse, enabling the handler to cross the water and walk with the horse. The horses tend to swim on their sides, making loud snorts to clear the water from their nostrils. After the assigned number of laps are finished, the bridge is drawn up and the horse clambers back up the ramp. The animal is bathed off with clear water to rinse the chlorine from his coat and covered with a blanket to keep him warm until he dries. It is kept toasty warm in the pool area. The wet horses are hung on the hotwalker to dry off, reminding me of damp clothes on a line. When all the horses are finished swimming for the day, the pool is vacuumed to remove their manure.

Wilbur's solution for dealing with those he doesn't like is to farm them out. Cindy, one of his former employees, now manages a little farm a few minutes away and is short of help. When she asks for the loan of some help, he sends over his least liked employees, including me, of course.

The ten acre farm that Cindy manages has a staff house, an indoor eighth mile track, a five stall barn and a larger main barn. The farm is compact and Cindy runs it well. Her regular girl Lisa is on sick leave and Cindy has no one to cover Lisa's work.

Cindy is exactly my age and is a graduate of a similar horsemanship course to the one I took at Humber. We get along from the first day. She enjoys the fact that she can trust me to run the farm on her day off and finish the to do list left in the tack room.

At Christmas, Wilbur informs me he is moving me permanently over to Cindy's farm in the new year. I am as relieved to be rid of Wilbur as he is of me.

In January, the racehorses go back in training. Lisa has recovered from her illness but on Saturdays, when Cindy has her day off, I have to pick up Sally at Wilbur's farm. While Lisa rides, Sally and I do the farm chores together. Sally is a foul mouthed brat who tries to get out of as much work as possible. The first day we work together, she suggests we simply pick the stalls out. I am not falling for this crap again. If we pick the stalls today, they will be twice as bad tomorrow when Sally will not be here. No, we muck properly.

The racehorse trainer Red comes daily to watch the horses train, bringing his exercise rider Jackie to help gallop. Jackie is surprised to see me mucking. "Why aren't you galloping?" he asks. I've known Jackie for years and he is quite familiar with my riding. I explain how, as the temporary winter help, I am **low man on the totem pole** and end up mucking while the other girls ride.

The owner of this farm has generously paid for Cindy and Lisa to attend a weekend seminar in February. Cindy has guaranteed me the two days of riding along with Jackie but Sally will have to do extra mucking. Jackie promises to put in a good word for me to Red and hopefully I will land a job galloping for him at the track.

Sally, of course, knows this. This was pure spite on her part to quit the very weekend that I have a chance to gallop. Even though she is jealous of me, I am so shocked by her callous back stabbing that I cannot even express my rage. Besides, I am not about to give her the satisfaction of knowing how much this hurts. If I cannot get on the horses, Red will not see me ride and hire me. I need to do something or I will lose this opportunity.

As I drive to the farm, I form my plan. At the farm, I go into overdrive. I race through feeding and trot the broodmares out to their paddocks. I cheerily wave good bye to the girls as they drive off and I am pulling the tractor into the barn. With seventeen large straw-bedded stalls to muck, there is no time for niceties. I smack horses that are not quick enough to move over. Usually it takes fifteen minutes to muck a straw stall but this morning, by hurrying desperately, I cut the time in half. By the time the guys show up at ten o'clock, I have mucked all the last stalls.

Red is amazed at how much I have accomplished. We go for coffee, then ride two sets. Red is also impressed with my riding. "How come I've never seen this girl gallop?" he announces to the world. "How come I don't know her?"

Jackie rides another horse while I have some lunch, then we ride one more set. Jackie is as good as his word and put his two cents worth in for me. Red sounds like he might hire me. When the guys leave, exhaustion sets in. I still have seventeen stalls to bed, bring in all the horses outside and feed.

When I wake up the next day, every muscle in my body aches. Cindy has learned how I was short staffed the previous day. She is up early to feed and turn out, easing my workload. She even mucks a couple of stalls before she leaves. Thank God. I am so stiff I can hardly lift a fork.

Before I am finished, Red and Jackie show up. Jackie starts riding while I am still mucking but I manage to do some riding. And it is official. Red has hired me to ride for him at the track. I will start when the horses go in just a week away. I am wildly elated.

Chapter Twenty-Five

RED MILLAR'S CAT

I can't say Jackie didn't warn me. "Red trains every day unless the track is closed," he told me. Despite the harshness of Ontario's winter storms, Woodbine training track rarely closes after training begins February 1. I am willing to brave the weather for a chance to gallop again.

In these winter mornings, my car has barely warmed by the time I drive through the backstretch gates. I park near the manure bins and shiver as I dash into the barn. Inside the tack room, it is cosy and warm. Red already sits in the only chair, holding court, retelling stories. The coffee pot beckons. The rest of us are scattered about on upturned buckets or old shipping trunks, nursing our coffees. The aroma of Bert's pipe adds to the homey atmosphere. I pour a cup of that awful brew and look for a place to sit. I feel very much at home in this barn.

In the corner, on a pile of horse blankets snuggles Moo, a calico feline. She barely glances at me. Secure in her status as barn matriarch, she is not about to give up her toasty spot. I find an used bucket to squat on.

Red is a true racetrack character. He is THE BOSS and runs his stable with an iron fist. With his cigar clamped in his bulldog jaws,

Red resembles Winston Churchill saving England from the Nazis. Red takes pride in being a grizzled ex-wrestler who is not afraid to pull a rider off his horse and boot him in the backside. "They call me Kick Ass Millar," he brags to all and sundry. The racing game is about winning and he is not above cheating, if necessary. The old man knows more about racing than anyone I have met.

On the other hand, Red can be a courtly gentleman, willing to help out a comrade or dispense advice. He habitually wears a fedora as a badge of his professional status. The way Red dispenses advice reminds me of the trainer played by Mickey Rooney in the movie, "The Black Stallion". His loyal crew rides out his tantrums and have been with him for years.

When Jackie finally shows up in the morning, Red says, "Let's get started." The first set is arranged and I select my tack and drop it by the stall for the groom. Half a cup of coffee later and I am on my way out to the track.

Mercifully, the first set is usually a pair of two year olds. Being unraced, they are easier to gallop and rarely pull. I bury my thinly gloved hands in Nesbit's neck and bow my head, letting the peak of my helmet shelter my eyes from the wind. After a mile and a half, Jackie and I jog half way home and escape to the tack room for some warmth between sets.

Credit: *Michael Burns*

Red is a true race track character. "They call me Kick Ass Millar," he brags.

While I was still on the farm, Red put me on a huge horse named Briar's Boy. He isn't the tallest horse I have ever been on but is certainly the widest. He is like sitting on a table, his back is so wide and flat. I sense the strength in this big horse but he rides nicely for me.

Red tries me on him the first day at the track. I find him equally polite to ride here. A week later, Red is eager to ramp up the training of the older horses. He asks me if I think I can gallop him along.

Galloping along is somewhere between a regular gallop and working. It is opening up the pace a notch and still controlling the speed. I gallop the horse around to the quarter pole as instructed, then lean forward, shorten my hold on the reins and cluck to this horse.

Briar responds by switching gears and flattens out into full work mode. I ask him to ease back but there is absolutely no response. When my efforts prove futile, I do the only possible thing. I fold up and ride the work out.

Privately, I am worried I will not be able to stop this freight train. As we flash under the wire, I stand up to signal that the work is over. A dozen strides later, I feel him begin to back off. I heave a sigh of relief that he pulls up nicely and walks home quietly. Now I have to face Red.

"The clockers got you in thirty seven and change," Red informs me as I ride into the barn. "I'm just glad you had sense enough to sit down and not pull him from side to side, fighting him all the way down the track." Red turns his back on me and stalks down the shedrow.

Jackie consoles me. "Nobody can gallop that horse along and Red knows that. He just wanted to get a work into the horse." However, Red never puts me on the horse again.

It is difficult to dress appropriately for galloping. Clothes must be roomy enough to move easily, break the wind, protect the extremities and yet not make you sweat. Once on horseback, nothing can be removed, lest your horse shy. Goggles are useless in driving snow. Mid length boots allows the skin to be chafed off calves. Exposed flesh is subject to frost bite. Thick gloves let reins slip but reaching for another hold causes your horse to run away. Most of us use the cheap knitted gloves sold at the track's tack store with some people not using any gloves. We are so bundled up that when anyone calls, "Hi, Jan," I have no idea who it is. Sometimes I can figure who it is by the saddle towel of their stable but lots of times, their identity remains a mystery. At this time of the year, galloping a mile on a good horse is merely a hardship. On a puller, it is just short of murder.

Once a horse starts to get tough, you have no choice but to fix your **crossed reins** against the neck and brace them there. You lean against the pressure, using your arms and legs for leverage. The taut muscles interferes with the circulation to your hands and feet, resulting in numb limbs. Up to a mile, I can give-and-take with the reins, allowing blood to flow into my arms for a few seconds. I pace myself, concentrate on breathing evenly, prolonging my strength. By the mile mark, all sensation is gone from my frozen fingers. If Legacy, one of the horses I gallop, continues to lean on the bit, the reins will slowly slide through my hands. On occasion, Legacy runs off for an eighth of a mile.

No matter how the horse gallops, I pull up at the out gap and begin walking home. I pry one hand off the reins and tuck it into a pocket to let my hand thaw. In a few minutes, the chilblains start.

As the constricted blood vessels open again, the red stuff rushes back in. It feels like liquid fire. Grown men have been known to cry if their goggles are down to hide their tears. Slicing my finger to the bone with binder twine was less agonizing than a bad chillblain. By the time I reach the barn, my hands might return to normal.

Instrumental is a particularly hellish ride. Jackie has given the filly the moniker Dizzy Izzy. She really is a 'head case'. In other words, she is nuts. She is a small, dark bay filly who can be a tough as nails. Jackie gives me specific instructions on galloping her.

"Drop your **pedals** right down. Take one hold and don't move. She'll pull from the **3/8 pole** but will ease up after the wire." He has already put a **German martingale** on her.

The German martingale consists of two short straps with snaps on either end. D rings are sewn onto reins. One end of the strap hooks to a **neck yoke** at the chest, passes through the bit and hooks onto one of the D rings. If the horse carries its head in a normal position, the straps just sit there. If, however, the horse tries to run off by putting his nose in the air, the straps double your pull on the reins.

Sure enough, the filly starts kindly enough. For the first half mile, the reins are loose while I enjoy the ride. Then she takes hold of me. I lean back and brace against my feet, using my body weight to control her speed. There is no give-and-take with her. In the bitter winter weather, my hands are numb and my arms ache. Sometimes she eases up at the **wire,** sometimes she pulls for the other half mile. Then she stops, turns in and heads for home in a single move.

Unlike other horses, Instrumental never walks home. I am a helpless passenger as she jogs along, the reins looped through my claw-like frozen digits. I steer as if I am riding a bicycle, leaning into turns, using my reins like a steering wheel. Only when safely inside the barn does she relax into a walk. I guide her into the stall and slide off.

Her groom barks at me. "Quick, grab the saddle before she **ties up.** Hurry!"

I stand in the doorway, helpless, trying to choke back tears. "Uh, Don, I can't move my hands." I am still trying to pry them open.

The weather is the enemy. Every winter day is just endured. The wind sears through my gloves. It sucks the breath out of my body and my eyes water as I gallop into it. Wet, driving snow is the worst. It stings needle-like into my eyes but gums up the goggles if I pull them down. It soaks my gloves, my jeans and my long underwear underneath. When the temperature rises near the freezing mark, my job is almost bearable.

On the worst days, Jackie and I often compare our frost bitten cheeks like badges of honour as we ride home. On these raw days, there might be as few as five exercise riders on the track. Jackie and I will account for two of them. But then, as I said, Jackie warned me.

It is on a day such as this when I make my resolution. I am chilled to the bone and need a cup of hot coffee to warm my insides. I step into the tack room, letting the heat break over me. In the corner, disturbed by the draft, the barn cat Moo raises her head. She glances at me, stretches her paw wide and with a sigh, snuggles back into her blanket. That's when I make my decision. Next time around, I'm coming back as Red Millar's cat.

Chapter Twenty-Six

Galloping Lessons

"She planted me like an ear of corn," is Jackie's colourful pronouncement of being dropped off of Peggy, the chestnut filly he is sitting on. Jackie hasn't been dumped in several years, yet this filly is already gaining quite a reputation.

Noel, the exercise boy who broke Peggy and had forty years of riding under his belt, just as easily **bought his own real estate**. If fact, she has dropped everyone who has been on her yet. Peggy is as slippery as an eel to ride, requiring constant vigilance.

Racing is about to start at Greenwood. With both tracks open for trainng, Red is splitting the barn and taking the older horses down to Greenwood, leaving the two year olds, including Peggy, behind at Woodbine. Jackie is relocating with the older horses but because of Peggy's reputation, Red asks Jackie to gallop the filly on his way down to Greenwood. This arrangement only lasts a week because the older contingent of horses cannot train until Jackie arrives.

Therefore, Red decides to let Alison, our newest rider, start riding Peggy. Personally, I think this a poor choice, as Alison does not yet have much experience with youngsters. I also find Alison is unwilling to listen to advice from me. "Alison, keep a hold of the chicken strap. Don't trust this filly an inch." I am on another two year old, keeping her company. Alison continues to ignore me.

On the second day, disaster strikes. With Alison momentarily distracted by horses galloping by, Peggy flings herself backward. Alison flips over the filly's head. Free of her rider, Peggy wheels around, heading out the gap we had just ridden through onto the track. I can only watch helplessly. There are times I hate being right. Besides, I still have to train the filly I am on.

A loose horse is always dangerous. They can cause accidents with other horses, collide with cars and otherwise injure themselves. Grooms run out from the nearest barn to catch our filly. Peggy evades them, racing full out on the paved road. Then she slips, skids and spread eagles on the tarmac. When she springs to her feet, it is obvious that she is hurt and she is soon caught. The barn logo on her saddle towel is recognized and she is returned to our barn.

By the time I ride back into our barn, Peggy is in her stall. Her injuries are serious. She has sliced off the skin from the front of her back pasterns almost to the bone. Gravel from the road is embedded in the wounds. Her martingale has ripped the tender flesh behind her elbow. She is a mess. Peggy's groom Bert gently washes out the deep lacerations with soap and covers them with a healing cream.

When Red checks in at the barn later, he is furious. The damage to Peggy's pasterns is so extensive, there is a chance she will never race. Alison, of course, feels terrible. Red relocates her to Greenwood where he can keep an eye on her.

For three days, Peggy is only hand walked while her flesh wounds scab over. On the fourth day, however, Bert tacks her up and motions me to mount up and take her to the training track. Because of where her skin is ripped inside her elbow, Bert cannot put a martingale on her now.

I am taking no chances. I arm myself with my stick, my pedals are dropped down and I have a hold of her mane. I also trot her out near the outside rail. Sure enough, at the spot where Peggy had dumped Alison, the filly drops her head and goes into a bucking spree. With my right rein, I crank her head around and wail on her with my stick. Around we go in such a tight circle that the filly cannot use her strength against me.

The outrider, who watches this escapade, is unimpressed with me. He is concerned for the safety of other riders passing by and bans us from the training track.

Next morning, I find myself banished to the back field at the far end of Woodbine. I have never ridden here. There is a tiny strip of dirt, about a mile in length, which is plowed around the edge of the field. This unsupervised track does not get much use.

It is lovely to ride in the back field. Because it is not policed, riders can work either way on the narrow dirt track. This makes it an ideal setting for reschooling Peggy and gives me some space to deal with her shenanigans. Horses gallop up behind us and pass us. Occasionally we meet a rider training the wrong way on the track. Any of these excuses can start a bucking spree. But now I am too angry to fall off. Each time I catch her braking suddenly, dropping her head or trying to duck out, I wheel her around, nose to tail, kicking her and making her spin. When her efforts to drop me prove fruitless, she gradually gives up her antics. The long walk there and back teaches the hot filly to relax. Still, Bert breathes a sigh of relief each time I return safely back to the barn.

One morning I happen on some Canada geese nesting out there. I can also see an approaching rider, galloping the wrong way. It is like watching a cartoon strip unspool. There's the rider approaching a little knoll on the right, goose on the left. Next scene has surprised horse splay legged in front of goose. Horse makes U turn. Horse goes off in one direction, goose in the opposite with the rider walking away from the scene.

It is a week before Peggy gives up her antics completely. I ride her a few more days out back before I return to the training track. She is fully healed, her lesson has been learned and she gallops like a princess. As far as I know, she never drops anyone again. I am pleased that my efforts to turn this filly around have been successful, when more experienced riders had not reformed her.

I have a lot to learn about galloping. Each rider usually has a roster of horses to train each morning. Northern Legacy is often my first ride of the day.

I first met the leggy three year old at the farm. He was on a six month lay up for a broken **coffin** bone in his foot. Cindy housed the gelding right next to the tack room where she and Lisa made a regular pet out of him. With his stall rest finished, he went into training along with the other horses in January.

In the first month at Woodbine, Jackie and I usually walk out to the training track together, then Jackie gallops off alone. I wait to put some distance between us, then start galloping Legacy. My mount gallops high headed and goes in **draw reins** until he develops sore shoulders after a month. Red listens to my input and we take the draw reins off him. I make a mental note of how detrimental draw reins are to riding horses. I also like the fact that Red sometimes trusts my judgement enough to listen to me.

Every day is different at the track. One morning, when it is still dark, Legacy spooks at something beside the track. I look over the rail to see a little red fox quizzically regarding us. The fox then trots off to the infield to see if the Canada geese have laid their eggs yet. I love how nature touches me even in the city.

At the end of our gallop, when I pull up, Jackie and I usually walk home together. Jackie uses this time to inform me about Red. He has sized our trainer up to a T. Often, Jackie and I will jog the first part of the way home. "Red can't see this far," Jackie explains. To prove it, Jackie makes a rude gesture in the air.

Then abruptly, Jackie pulls up. "Okay, Red can see us now." Reins relaxed, Jackie and I nonchalantly stroll the rest of the way home.

However, if Jackie screws up, he cuts through the nearest barn, jogs down their shedrow and over to our barn. Here he makes a quick change of horses before the boss can march back to thoroughly chew Jackie out. By the time Jackie finishes with the next horse, Red has cooled down and perhaps even forgotten what he was been so mad about. I don't get so lucky. As the new kid on the block, I have to take it on the chin. Red never holds a grudge and he never takes me off a horse I can handle. I begin to learn to finesse the trickier racehorses and am given more opportunities. Jackie is a generous teacher, always looking out for his less experienced fellow riders. I am lucky to be paired with him. I never, however, work any horses in our barn. Usually the jockeys do the workouts but Jackie, who was a former jockey, handles some of the them.

Someone tells me Jackie's back story. As a jockey, Jackie went racing State side and was saving his money to retire. He almost had enough to come home when he was busted up in a race. With no insurance, the medical bills ate through the money. Jackie was broke by the time he was healed enough to come home. Now he gallops in the morning and works a second job in the pari-mutuels (betting windows) in the afternoon. I am forewarned that riding racehorses comes with no guarantee of safety.

In May, racing shifts to Woodbine and our older horses come back from Greenwood. The weather warms and we dig out our T shirts. Galloping abruptly changes from brutal endurance to the best job on the track.

Every morning, the training track is closed for half an hour for harrowing and we schedule our galloping to take this into account. Shortly after the barn is united, one of the grooms says, "Quick, before they close the track, can you jump on Instrumental? She's all tacked up, ready to go."

I had been glad when Instrumental shipped to Greenwood and I did not have to gallop her for two months. She really is tough to ride and pulls like a train. Now I leap off one horse, run down the aisle and jump on the filly. I notice that the German reins are on the loosest ring, not the tighter ring that I used. Since Alison has helped tack up the filly and has been galloping her, I assume Instrumental is fine to ride this way. What I do not know is that the filly has run off the last two times Alison galloped her.

Within seconds of hitting the track, I realize I am in trouble. The filly pulls right from the start. With the German lines so loose, they are not helping to control her. For a full mile, the filly leans into the bit. I can feel my strength ebbing.

My one hope is that I can yell for the outrider as I ride past and he can intercept us. That hope dies when I see the outrider standing beside his pony, gossiping with onlookers.

Just past the wire, my hands go completely numb and the reins slip. With the tension on the reins eased, the filly bolts. By sitting on the filly with my pedals long, I have no leverage and no hope of controlling her speed. Instrumental kicks into racing gear. We are flying around the track. I pray she will ease up on the backstretch where we usually pull up but there are too many riders there. By the time I can pull onto the outside rail, we are well past that point. The filly has no intention of stopping now.

Unable to pull up, I have no choice but to steer her back into the middle of the track. After a full mile of running, she finally runs out of steam. When I get her onto the backstretch again, she practically stops of her own accord. She is dead quiet to walk home.

Red is, understandably, pretty upset. He does not blame me, seeing as the filly had run off with Alison at Greenwood until the outrider had caught her. I learn now the filly is already sold. To keep her safe, Instrumental is hand walked until she leaves a few days later. We all breathe a sigh of relief when the filly ships out.

Jackie is going to work Briar's Boy today. This is the brawny chestnut gelding I only galloped for a week. Since we are not that busy, several of us head over to the track to watch. Jackie has his **pedals** hiked all the way up but dangles his feet out of the stirrups as he walks the big strapping horse out to the track. When Jackie comes by us, he is still galloping with no stirrups. He waves the knot in the reins at us, to show off how easily this horse rides for him. Jackie has taught this horse to hack along if Jackie sits down on the animal's back. He could be riding some kid's pony, the way the horse is going. Jackie banters at us and we kid him back.

As he approaches the half mile pole on his second trip, Jackie lifts his feet and quickly slips them into the stirrups. The gelding immediately accelerates into high gear. Jackie is now bent completely in half, bracing against his feet that are pushed out in front of him. We can see the strain on Jackie's face as he struggles to control the speed.

"You got him, jock!" we shout at Jackie. It is not fun for Jackie now.

Another two year old ships into our barn now. He's a little bay gelding and I mean little, just 14.2 hands high. By definition, he is actually a pony.

He's a fun guy to ride and all three of us take turns galloping him. The consensus among us is the pony has no talent.

When the owner shows up to watch his pony work, the three of us ride out in a set together. When we get to the track, Jackie, who is riding the pony, gives us our instructions.

"We are going to work from the quarter pole this morning and I am going to finish on top."

Jackie positions himself between Alison and me, with Alison on the outside. I feel sorry for Alison. Woody, her mount, is a full hand taller and much stronger than Nesbit, my mount. Nevertheless, when we get to the quarter pole, off we go.

By hiding Jackie in the middle, the owner cannot see Jackie beating on the little horse for all he is worth. Alison and I are standing in our stirrups, strangling our mounts. Sure enough, by the wire, the pony is half a length ahead. We pull up together and walk home. Red is pleased with us. The owner is thoroughly convinced of the worth of his little horse. I am disgusted by the con job on the owner.

The pony races twice. Both times, he is **outdistanced.** Finally, the owner takes the horse home.

One afternoon, I show up to our barn to learn that earlier in the day, Legacy broke his leg in his race and was put down. The previous injury to his coffin bone was a ticking time bomb. The horse was a real favourite of mine and I am heartsick. At the end of the race day, all the races are replayed on the in house monitor. I go down to the kitchen to watch his final race.

Legacy looked like an easy winner, pulling away from the field. Then the horse bobbles and the jockey stands us in the stirrups to pull the horse up. The field sweeps by and the camera loses sight of the broken down horse.

I cannot imagine the grief of Cindy and Lisa at the farm. They made such a pet out of this sweet horse when he was cooped up in a stall for months on end. I remember how lovely he was to gallop and how he spooked at the fox months before. If only there was some way to know if your horse has a potentially fatal weakness. I know the horse was sound when we sent him to that race. I'm glad the jockey wasn't injured but my heart aches for this lovely horse.

As the two year olds start racing, there is less and less riding for the exercise people. Whereas I might ride seven or eight horses every morning in February, by summer this number is cut in half. Red has already let Alison go so it is no surprise to me when I am laid off in June.

Chapter Twenty-Seven

HEADING SOUTH

No matter how careful I am to stretch every penny, there is never any money left after I pay my essential bills. I am tired of being poor. I have to earn more money if I am to change my circumstances in life.

While I am galloping, the shorter work day does offer me the chance to earn money elsewhere but I have few other employable skills. I do find freelance work around Woodbine - horses to run or hots to walk after a race. It doesn't pay much. Five dollars to cool out a horse, ten to run one. I only find a couple a week to walk and rarely, a horse to run. This amount of extra work certainly isn't filling my bank account.

Fortunately, I pick up a steady client in the next barn. Shelley started at Woodbine twenty years ago but moved to Maryland with her trainer and stayed there. She worked as assistant trainer, then became a trainer in her own right. Shelley relocated back to Canada just a few months ago. Her groom is pretty much the dregs of the barrel and I help her out on several occasions. When Red lays me off, Shelley offers me a full time job and fires her previous groom.

We get along well. She rubs a couple and I have four to groom. They are the worst behaved animals I have ever dealt with. None of them kindly hold their feet or move over when told. Queenie nearly flips the blacksmith over when he picks up a hind foot. She is such a pig for the pony boy that we are forced to put a chain through her mouth, then back over her nose in order that he can control her. This is an incredibly severe restraint.

Shelley does most of the galloping but sometimes I ride out with her in a set. I have lost some weight this year and am quite a bit thinner than Shelley so she has no issue with my weight. I am still not thin enough for many of the trainers to hire me.

Shelley and I walk hots as needed. I get one day off, otherwise, I feed every afternoon.

On Shelley's day off, I gallop her little black horse Chappie. Chappie belongs to Shelley and she has thoroughly spoiled him. But he is funny. While walking to or from the track, something will catch his eye and he'll stop dead and stare. Five seconds later, he'll swing his head back and forth, squeal and leap straight up in the air. Shelley laughs every time he does it and does nothing to reprimand him. To her, it's harmless fun.

Another weird thing about Chappie is that he is terrified of anything white. If I have to muck his stall with him in it, I must close his screen before sprinkling the lime dust we use to dry the stalls. Chappie bounces off the wall, trying to avoid the monster dust. Once I pull down the straw and cover the lime, he relaxes.

The day comes when I have to put an Uptite poultice on his front legs. Every barn uses gallons of the white clay to make wet casts, which are then wrapped in plastic and a bandage.

I manage to do Chappie's left leg without incident. Once I have the clay on his right leg, however, he looks down to find his leg has turned white. Terrified, he tries to get away from his leg while he is still tied to the wall. I manage to undo him and let him freak out until he realizes he can't outrun his own leg.

Two of my fillies are really sweet. I make arrangements for a couple of childhood friends to come to the track to watch the training. Shelley signs them in. With their red Guest badges on, my friends take in every detail of track life. Shelley and I ride a set, allowing my friends to watch me gallop. Afterwards, I get permission from Shelley to let the friends groom the two quiet fillies. Both girls own their own horses and can handle these fillies. They are thrilled to be a part of my day. I rarely get the chance to share my passion with friends from outside the track.

One of the fillies is a cribber. **Cribbing** is a learned behaviour, usually stemming from boredom (often because an injured animal is stall bound) in which the horse grabs a ledge or other handy item, flexes its neck and sucks in air, making a characteristic burping noise. Cribbing is obnoxious for a variety of reasons. The horse damages his teeth, tears fences apart and the habit can cause colic. The noise of cribbing is a habit I can't stand, akin to scratching fingernails across a chalkboard.

We have my cribber entered in a race. The night before her race, when I come in to feed, the filly is laying down. I get her up but within minutes, she lies down again and refuses to eat. I check my watch. It is now four thirty. I feed the others and tidy up but I have a bad feeling about this filly. I hang around and keep an eye on her.

At six o'clock, she finally begins to colic. I call Shelley. Shelley drives in but we have to treat the filly and get her scratched from her race. Within minutes of being dosed, she is up and eating her dinner as if she never had a bellyache. We attribute the colic to her cribbing. It won't be her last colic episode, either. The needs of the horses are paramount and I resent these intrusions upon my time. I don't have the easy going personality to be a great groom.

Shelley has decided to take some horses to Maryland for the winter and asks me if I want to go with her. I have had a hankering to work in the States and this provides me with a legal way to do so. I can live on the track and keep my apartment in Toronto. Shelley leaves me to load the horses on the transport van and she drives down the day before me. I will follow the van with my car.

The van is supposed to be here at seven a.m. At nine o'clock, there is an announcement over the P.A. that the driver will be three hours late. I use the opportunity to walk each of the horses and let them stretch their legs. In hindsight, I should have gassed up my car instead.

The driver shows up at eleven. He is doing back-to-back runs and needed to stop and get some sleep. The horses load without incident and I follow the driver onto the highway. I watch my gas gauge nervously. As we get close to the border, I decide I must stop for gas. I pull off the highway, believing I will catch up to our horses at the border. What I don't know is that the border crossing for livestock has changed since the last time I crossed with horses. I bypass Niagara to cross at Buffalo and totally lose track of our van.

I enter the New York Freeway and I drive an hour past my exit before I realize my map is wrong. I turn around and retrace my route until I can head south. By seven pm, when I stop for supper, I have just reached the end of the throughway. The radio has been calling for a snowstorm to start about midnight. As I finish my meal, the first of the flakes begins to drift down.

This becomes the most nightmarish drive of my life. It becomes a test of endurance. The freeway ends and narrows to a two lane highway that winds through every little town, all the way through Pennsylvania. As the snow piles up, semis crawl up hills at ten miles per hour. There is no room to pass. I blindly follow the cars in front, unsure of where I am. I am scared to continue but even more terrified to stop.

My eyes ache with the strain of steering into the driving snow. Since I do not have a credit card and only limited cash, I cannot book into a hotel. Besides, no one knows where I am. I was supposed to be in Maryland hours ago. I have no way of contacting Shelley since I have neither a credit card of my own nor a long distance phone card. God knows where the horses are.

By eleven pm, exhausted, I pull off at a fast food restaurant and sleep for a couple of hours. The cold seeping into the car wakes me up. When I resume driving, the traffic is much reduced. I stop to refuel but within a few miles my car starts to chug ominously. I recognize that my fuel line is freezing, that there must have been water in the gasoline. I find another gas bar open and buy gas line

Heading South

anti-freeze. In seconds, my car purrs smoothly again. One problem solved.

I get lost once more, in Baltimore before pulling up to the stable gate at seven a.m. I learn now that my horses have not even arrived. I phone the contact number Shelley has given me. She picks me up and puts me straight into bed. Having been up over 24 hours, I fall asleep immediately. When the horses come in later, she unloads them and feeds for me. It is late afternoon before I wake up.

Next morning, I arrive at the track to find several of our horses running a fever. They were on the trailer for close to thirty six hours. The driver had pulled over to let the storm blow over. I walk the horses while Shelley mucks their stalls. Then she sets me up with a tack room to live in while I am in Maryland.

Everything about this track is so different. My comfortable room is large, private and warm unlike the diminutive tack rooms at Woodbine which open directly onto the outside. I have seen snow drifts in those rooms. I am not forced to share a room with a stranger either. I am quite happy with my simply furnished room here.

There is no training track here at Laurel track. Even the main track is small. Every square inch of the backstretch is pressed into use. Between the barns, rubber mats are laid over cement and hot walking machines are set up. The hot walking machines are more common here. The exercise riders are much heavier, too. Many of them must weigh at least one hundred and fifty or sixty pounds. In Toronto, as a girl, if you weigh any more than one twenty-five, you can forget getting on horses. Only the most experienced exercise boys willing to get on the roughest horses can be a little heavier.

I don't get a full day off here but I don't miss it. I find plenty of things to do see and do within a short drive. There is a huge mall nearby, blocks long, which is even open on Sundays. I begin making friends among the grooms and feeling settled.

Unlike Ontario, which only has three racetracks that make up our circuit, there are numerous tracks in the area which are open for training. We are stabled at Laurel track but ship out to race. One afternoon, after racing, I almost get on the wrong transfer van, going to a different track in a different state! The driver fortunately catches my mistake. I cannot imagine how I would have solved the dilemma of winding up on the wrong track. We are busy running our horses and they all pick up cheques. A couple of them are claimed, which was part of the plan of shipping down here.

After three weeks, Shelley comes and asks me whether I would like to go home. I've seen enough of the States and miss my family. Shelley lays me off and I quickly pack up to head back to Canada.

The trip is so different to the one coming down. The weather is fine, the drive straight forward and I stop a couple of times to shop. I reach the border late in the night and am whisked through. I head straight to my parents, who are over the moon at my arrival. I have been gone just three weeks. It is the wee hours of December twenty-fourth. How wonderful to be home for Christmas.

I am glad I went and delighted to be back. I've learned a lot about my ability to cope with unfamiliar situations and strengthened my confidence in myself. I am building more than horse skills; I am developing the self esteem I need to fashion my future.

Chapter Twenty-Eight

A Brand New Farm

As soon as I get home, I call up Cindy, where I worked last winter to see if they need any help. She welcomes me back. It is so rare for me to return to work at a place I have once been but I like working at this farm.

The previous year, the owner of this farm bought fifty acres of farmland, planning to upgrade his breeding operation to a bigger facility. Now the new farm is finished, the old farm is sold and we have to move everything, lock, stock and barrel. Just after the new year, we load the horses on vans and ship them over. All the tack and farm implements are stuffed into the loft of the vans. Then the remaining hay is stacked on the farm's half ton truck. By the end of the day, everything is already neatly put away at the new farm.

The new farm is marvelously laid out. Because the paddocks fan out from the barn, no horse is led more than fifty feet from the barn. The fencing is post and rail, beautiful, safe and expensive. There are two houses on the new property – an older mansion completely renovated for the farm owner and a freshly built house for the staff. The centrepiece of the complex is a huge indoor arena for conditioning. The main barn is built at one end of the arena. In the front of the arena, half way down the long side, is the office.

The front part of the office is the domain of the owner who has consolidated his business onto the farm. Behind, our lunch room has a picture window which allows Red to watch the training of the racehorses in comfort. Past the office, attached to the other end of the arena is the broodmare barn. Now, in foul weather, we can cut through the arena and not walk outside except for turnout of the horses.

We quickly settle into our routine of feeding, turnout, mucking and training. By returning to the farm this year, my status has changed. I am not brushed off as temporary help and my opinion counts more. It is a given that I will return to work for Red in the spring. I take a more active interest in the farm, too. I find ways to implement some labour and time saving routines. The farm hires only one additional staff, a boy to help with the mucking and other farm chores. I am no longer the 'low man on the totem pole'.

There have been other changes since I last worked here. The one mare I galloped early on at the track came home to be bred and is now heavily in foal. The weanlings from my first work stint are now broke and are the young horses in training. Nesbit, the two year old I loved galloping last year, is now a broken-down cripple and will make someone a pleasant hack horse, if he comes sound enough.

Shortly after moving in, Red buys a good looking gray filly and ships her to us. She is a sturdily built two year old and Red has high hopes for her. The first night, she digs an impressive hole in her stall about two feet deep. The barn isn't even two weeks old yet! The next night, she does it again. Cindy moves her into a stall with a rubber mat lining the floor. That should fix the filly's little red wagon!

A couple of days later, I am the first one to the barn and start feeding. When I get to the gray filly's stall, she is lying on her side, **cast** against the wall. A horse becomes cast when they roll against the side of the stall with their legs folded against them. With young thoroughbreds, it is a fairly common occurrence. Usually they can be pulled away from the wall or rolled over and the horses will scramble to their feet.

This filly is different. When she tries to get up, she cannot coordinate her back end. She is unable to rise. Panicked, I phone Cindy and catch her just as she is leaving the house. She races down to the barn after calling the vet.

Red soon shows up. Once the vet arrives, we are able to get the filly to her feet but she definitely is not right. She seems semi-paralyzed, unable to walk correctly. We guess that the filly has laid down and then slipped getting up on the wet floor mat. She must have twisted her back. The decision is made to just keep her quiet.

Believe it or not, the filly lays down the next night and cannot get up again. This time the decision is changed to sending her to the equine hospital at Guelph. I am surprised we even can load the poor thing.

Several weeks later, she is shipped back to us, somewhat stabilized. She has developed a huge **hematoma** on her loins, where her back joined her hind quarters. Because of her digging, Cindy is reluctant to put her in a stall with a dirt floor. To prevent her from slipping on the rubber mat, we let the stall develop into a **deep litter bed.** We did not muck it, just pick out the obvious manure and let the wet straw build up underneath her. It works. Soon she is able to lay down overnight and get back onto her feet by herself. The filly has no turnout at this time but she is slowly improving.

The move to the new farm isn't without hiccups. A couple of weeks after moving to the new farm, Cindy's elderly dog disappears. Cindy is completely attached to this dog. Lisa, the other full time girl, kids her that when the dog dies, Lisa is going to have the dog stuffed and mounted on rollers to convince Cindy the dog is still alive. This, however, is no laughing matter. It's the coldest day of the year. Cindy fears the dog is confused and has started back to the previous farm, some ten miles away.

As animal lovers, there is no question the dog's well being trumps our work. Both Cindy and Lisa go looking for the dog, leaving me to mind the farm. Finally, in midafternoon, I get a phone call from the vets. Someone has found the animal curled up in a snow bank. They picked him up and traced the dog through the tag on his collar. I have to wait until Cindy returns to let her know where to fetch her dog. Sure enough, the dog traveled more than five miles toward the old farm.

Training has started in earnest. Six days a week, Red shows up to observe the horses' conditioning. Coffee cup in hand, he watches from the lunch room. The girls do most of the training but I now get to ride on each girls' day off.

One of the babies was not backed in the fall due to her small size. The first few days under saddle are the most important of a riding horse's life. Cindy has the day off and asks me to back the filly. Lisa is to assist me as ground man.

I **belly** the filly in her stall, after removing the feed tub and water bucket. The filly is fine to lie over and obediently walks in circles as Lisa leads her around. However, when I swing into the saddle and try to ride her in the same circles, the filly rears up, pinning me against the stall wall. Lisa freezes, completely forgetting how to help me. I am in the most dangerous of situations and Lisa is unable to react. When the filly comes down, I am able to spin her around myself. I am angry at Lisa for leaving me in such a bad spot but keep my mouth shut. She didn't abandon me on purpose.

The next day, I substitute the boy instead of Lisa. He is fearless, capable of following my instructions and proves to be a wonderful ground man. I would rather risk my neck with an untrained person that someone who panics.

At Woodbine, the grounds crew have been watering and harrowing the training track around the clock to soften it. In just a few weeks, it is ready for training. We get the green light and load the racehorses up to ship in to the track. I am delighted to be returning with Red.

Chapter Twenty-Nine

Proving Myself

There are a number of changes this year at Red's barn. For one thing, Jackie is gone. Red is getting crustier in his old age and Jackie is just the first to leave. We hire a young kid named Paul who wants to become a jockey. Red has the reputation of giving deserving kids a chance to race ride (become a jockey), putting Red into a minority of trainers on the track. Paul is a fairly useless exercise boy but later becomes a successful jockey.

When Greenwood opens, Red splits the barn as usual. He takes the older racehorses, the girls who rub them and Paul to move down where the racing is taking place. He leaves Bert in charge of our contingent at Woodbine, with Charlie hot walking and me galloping the youngsters.

With extra stalls available, a couple of other horses ship in. One is New Gold. He is a strapping chestnut gelding, built like a brick shit house. He is three years old and so far has made a measly two hundred and thirty five dollars as a two year old. Red has inherited him off a young trainer. All we know about him is that he gallops in blinkers. That's not much to go on.

The ever cautious Bert tacks him up, boosts me up and leads me all the way out to the training track. As Bert watches, I stroll along the outer edge of the track, trying to assess this new mount. I squeeze with my legs and ask for a trot. In response, the horse flattens his ears, stops dead and threatens me with a little buck. "Just try me," he seems to say.

Hmm. Not good. I try again with the same result. I try throwing the reins up his neck and cluck. Not exactly high tech but it works, although the cluck seems to annoy him. We manage a little trot. It is the same hoopla to get into a canter. The more you push, the less willing he is to go. Eventually, after a quarter of a mile, he consents to canter.

He settles into stride. He lacks the elasticity of a great horse but he is powerful and workmanlike. For half a mile, we gallop along easily. As we near the top end, he begins to pull. He leans into the bit until my back aches with the strain. My fingers go numb. Past the wire, he eases up of his own accord and gallops politely to the out gap. I pull up and we head home.

God, he is difficult to get home. He bitterly resents any directions from me and lets me know this in no uncertain terms. He jogs home, bucking whenever I try to rate him and kicks out at other horses that are passing us.

Bert meets me back near the in gap. As he snaps the lead line onto the bit, he says, "Better take a stick tomorrow, Jan."

"No stick, Bert. This horse is so sour, a stick will only make him worse."

The next day is a repeat of the first and the day after that is still the same. The only thing that changes is he begins pulling harder. We try an equipment change, putting the German martingale on him. This does the trick. When he begins to pull, the extra straps double my pressure and I am able to keep the chestnut gelding under control.

In our barn, the horse gains the nickname Goofy Newfie. It certainly suits him. Even riding down the shedrow after a gallop, I have to watch he doesn't kick someone. With his uncontrolled kicking, he can never be shedrowed. I am sent out to gallop in some pretty miserable weather. That's how I discover one of his talents. He gallops the same every day no matter what the condition of the track. It can be frozen, slippery, greasy or heavy; it makes no difference to him. It could be cut glass for all he cares. He is so sure footed that he can buck and leap all he wants and never makes a misstep.

His bucks are never very worrisome. It is his ability to kick out that is so different. He'll drop his head slightly, then fire both hind feet skyward. I find it easiest to stay off his back and balance on the **chicken strap** to avoid his shenanigans. He prefers to trot home without interference and I let him have his way. I get plenty of advice from the other exercise riders. "Pull up his head. Give him a smack on the bum." I ignore them. The less I interfere with Newfie, the better he trains.

Bit by bit, I manage to piece together Newfie's past. He was broke by an older exercise rider who would not put up with his nonsense. Newfie fought back, only to be hit again and again. When Newfie first came to the track, his young trainer had an inexperienced exercise boy that Newfie managed to terrify. Eventually, the kid rode in the in gap and out the out gap without ever completing a lap of the track. Somehow, Newfie managed a fourth place finish. That sums up his career.

Scottie is a whole different kettle of fish. Red buys the lithe red two year old out of sale in Florida on the strength of a promising workout in February. When the gelding is shipped north to Canada in midwinter, his sleek summer coat is no protection for our bitter winter. Scottie promptly catches cold. Cindy and Lisa should have been **legging** him up at the farm but because of his cold, they never ride the horse. The young horse is a completely unknown quantity when he is shipped to us at the track.

This first morning, cautious old Bert tacks the horse up in a simple **snaffle.** He calls me into the stall. "Since we don't know anything about him, Jan, I'll leg you up in here."

Bert lifts my calf and I spring up. I spend exactly one and a half seconds on Scottie before finding myself planted in the straw on the far side of the horse. Lucky for me we are not in the shedrow and I am unhurt. I decide to mount the colt as if he has never had a rider on before. I have Bert boost me up so that I lay across the saddle, then gently feel for my stirrup and swing up into the saddle. Scottie zooms around in the stall but he doesn't buck and I stay put.

Because this horse is acting so **green**, I trust my hard earned instincts and drop my pedals down. Bert leads me a turn of the shedrow as I tie on and then leads me out to the track.

About the only good thing I can say about this young horse is that he goes forward. Well, sort of. He acts like he has never galloped on a track. I settle him into the middle of the oval, giving myself plenty of room around us for safety. When horses gallop by us, he veers sideways as if hit by a gust of wind. Horses working on the rail cause him to throw his head up while trying to bolt. Fortunately, he still has the tender mouth of a baby and I am able to get him round.

Later that morning, Red returns from training at Greenwood. "How'd that new horse go?" he asks me hopefully.

"Red, that horse isn't even broke!" I fire back.

"What!!!" he growls in disbelief.

"He threw me off in the stall with a leg up. Then he galloped green as grass all the way round the track, spooking from every horse that went by him and he doesn't rate at all. He doesn't know anything."

"He has to be broke. I bought him out of a Horses of Racing Age sale in Florida and they have to show a work to get in that sale."

"Well, I don't know what they did with him in Florida but he isn't broke," I replied stubbornly.

Bert discovers that by facing Scottie into the wall when I am legged up on him, Scottie can not do that quick duck away that got me dropped that first morning. Scottie can only shuffle his feet and I land safely in the saddle. The problem is that Bert needs one hand to hold the horse and the other to give me a leg up. Since Bert is elderly and I am not the most athletic rider, Bert finds it too much of a strain to leg me up. He decides to use a freelance rider.

The first exercise boy lasts a couple of weeks before demanding more money. The second exercise rider doesn't even last that long. The horse is such a pain in the ass that the boys are not willing to be bothered. At least, not for the money that Red is willing to pay. So I inherit the gelding back.

He still whooshes sideways when horses blow by us, working on the rail. I learn to keep my head up and anticipate the trouble spots, moving him farther out from the rail to prevent us from interfering with anyone. Other than that, he trains fine and is soon fit.

Another horse I gallop while Red is at Greenwood is Fifi. That is just her barn name but it suits her perfectly. This rather rotund little bay filly is easy to gallop but the real fun is coming home. She is such a good feeling devil and will squeal, bounce up and down and hop from one back foot to the other. Fifi just cannot contain herself but she never tries to unload me. I laugh aloud at her antics. Other riders smile at us as they pass by.

Bert turns sixty-five and retires. We hire Fred to replace him.

Before racing returns to Woodbine, the officials close the training track for two weeks to recondition it. They open up the main track for training instead. The main track is a pleasant gallop. The track is much wider and the scenery more varied. To reach the track, we have to walk through a tunnel under the turf track. A crowd always gathers at the in gap on this track consisting of the **outrider, meat wagon** driver, various trainers and jockey's **agents.** I need Fred to lead me out to the middle of the track and start galloping from there, otherwise Newfie will simply run over the crowd.

Red has come to watch us train this morning. As Fred leads us out onto the track, Newfie begins to jump up and down and swing around Fred. We are facing the wrong way of the track when Fred lets us go. Newfie actually canters backwards for five strides before I can turn him around. Fred looks up at Red, sitting in the stands, and shrugs his shoulders, as if to say, "Had nothing to do with me".

Fred never waits to pick me up when I finish galloping Newfie. With Newfie going into his buck-and-kick routine in the tunnel, I am forced to yell "Heads up" and let everyone else part like the Red Sea. One time, in the tunnel, Newfie kicks so high his tail whomps me in the back. By balancing on the chicken strap and crossing my fingers, we manage to get home in one piece.

Red trusts me more this year. As Newfie gets fitter, Red asks me to let him gallop a **two minute lick** on him. Newfie is actually an easy horse to gallop along and **rates** nicely. To work him, though, Red brings in an experienced exercise boy. Newfie actually works nicely for him but I sense that he is happier when he is playing games with me.

When I learn that Red has Newfie entered for his first start of the season, I waylay the boss. "Red, Newfie hasn't been out of the gate this year."

"Oh, yes, he has," Red blusters at me. I know better than to contradict the Boss when he is in such an obstinate mood.

So when the gates open in his race and the rest of the field breaks alertly, Newfie literally walks out of the gate. He loses about twelve lengths at the start and is only beat a total of eight lengths at the finish. He would have won with a decent start.

Red has his poorest opening meet ever. He hasn't won a single race. He is still in a cantankerous mood when the barn is consolidated back at Woodbine for the start of the summer meet. A few days later, he puts me up on Snip and asks me to gallop him. Paul has been riding this three year old until now. When I ride back into the barn, Red meets me at the stall as I am unsaddling.

"Whatyathink?" Red inquires hopefully. Snip's sister was a **stakes** winner and great efforts are expected of Snip.

"Red, this horse isn't fit."

"Whatyamean, this horse isn't fit!" Red bellows. "He's run twice already."

Not very well, I think to myself. He hasn't even hit the board. I debate for a couple of seconds. "Red, do you remember when I galloped First Summer Day, Snip's sister last year?"

"Yeah," Red replies cautiously.

"And do you remember how she used to hop up and down all the way home? How I could hardly get her home?" I continue.

"Yeah-h."

"Well, Snip walked home on a loose rein, like an old riding horse."

Red digests this for a moment. "Okay, Jan, tomorrow you gallop him again. And take a whip with you."

Next morning I hop on Snip carrying the whip. When I ride out of the barn, Snip stops, turns his head and stares at the whip. He knows exactly what is up. Snip begins to put out. Instead of lollying around, Snip puts some effort into his gallops. Within days, he too begins to fidget about coming home, still bouncing with energy. Red is encouraged and, two weeks after I started galloping Snip, Red enters him into another race. Snip is bumped as he comes out of the gate but rallies to finish third. Next time out, he breaks his maiden. Then he finishes second in the next race.

One morning, Red decides to give Snip a work. Since I do not have a horse to gallop, I end up on the end of the shank, cooling out the colt. As Snip catches his breath and begins to cool down, he also begins to limp. I try to get Red's attention without alerting the owner. But as Red talks, the colt grows steadily worse. Finally, I have to interrupt the trainer.

"Uh, Red, Snip has bucked shins. Can I stop cooling him out?"

By now, Snip is not putting any weight on the sore leg, hopping painfully on his three good legs.

"That's not bucked shins," Red roars. I am instructed to put the horse in his stall. I hold the shank while Red picks up the foot and bends the knee. He carefully squeezes along the shin. Part way along, Snip reacts so badly he nearly falls to the ground. It is bucked shins.

Snip just has no pain tolerance. He is shipped home to the farm to let his leg heal. In my mind, I contrast him with Magpie, the little filly I rubbed my first year, who finished fourth with bucked shins. That filly sure had class.

Red buys a horse for himself, a nice chestnut gelding. We call the horse Red, as well. Although Paul usually gallops the horse, Red asks me to gallop the horse the day before his first race for us and send him along for a quarter of a mile. I am not sure whether this is because Red trusts my judgment or for some other obscure reason,

The Red horse seems a nice enough ride. His groom walks me out to the track and the horse starts off well. Along the backside, though, he takes a bad step and nearly falls down. Shaken by this unexpected misstep, I pick his head up but the horse continues along as if nothing has happened.

When asked for more speed, the gelding nicely moves out. Halfway to the wire, he bobbles again. My heart is in my mouth but nothing else happens. I pull up without incident and walk home.

Red meets me at the stall. "Will he win tomorrow?" he asks eagerly.

I hesitate. "He could. And it could be his last race."

When pressed for more details, I tell Red what had just happened. This horse is breaking down.

Red decides to go for the win. A **buzzer** is produced when they are prepping this horse for his race the following day. This is highly illegal. The assistant trainer zaps the colt repeatedly where the stick will hit him. This makes the horse put more effort out when hit by a stick, even if he is hurting. I am sickened by it all. The horse does win. Then Red ships him to the States. The horse runs twice more. Both times he breaks down. Too bad. He was a nice horse.

Newfie consistently **hits the board** but does not break his maiden. When Red goes down to the States for the race of his own colt, he leaves instructions that I am to gallop Newfie a two minute lick in preparation for his race the following day. I simply forget and gallop the usual two miles. When I ride back into the barn and am asked how he went, I realize nobody was watching. "Good," I reply.

The next day, Newfie romps home, breaking his maiden. It confirms my opinion that the horse trains better for me than outside riders. However, I don't dare admit my mistake to Red.

So Red continues to bring in outside riders before a race and Newfie continues to hit the board but not to win. I am frustrated, believing the horse will run better if I ride him but I keep my opinion to myself. For now.

Chapter Thirty

LAID UP AND LAID OFF

In the past couple of years, I have become increasingly frustrated working at the track. I am not making enough money to enjoy my life, the erratic hours make it difficult to work anywhere else but mostly I am frustrated by the lack of respect for the knowledge I have gained.

I have no interest in becoming a racehorse trainer. The boot licking, the running sore horses until they break down, the myriad of details that are out of your control, no, these are not for me. Besides, I have no interest in the races themselves. I just know I do not want to spend the rest of my life on the track.

If I stay on the track, I could end up like the sorry old men I see; too poor to retire, nowhere to call home, walking horses that are too strong for them until one morning, someone finds them dead in bed. I have seen this happen more times than I would like to admit.

Often, when I am away from the track, I still follow equine events. Both my sisters ride and I occasionally get out to watch their respective lessons. While watching them ride, I think to myself "I can teach a whole lot better than that drivel they are paying for."

That is how I figure out what I want to do. I want to become a riding instructor.

My trouble is, I have virtually no show experience and have only jumped a couple of courses in the past ten years. I know nothing about counting strides or putting a horse on the bit. My experience of teaching is limited to a couple of friends on The Old Man. On the plus side, my stable management skills are excellent and I have ridden dozens of horses in the past few years, developing a great eye for horses – lamenesses, their way of going. I just have to find a way to plug the gaps in my knowledge.

By looking at ads in a local magazine, I find a riding instructor who is already certified. I begin taking private lessons with her planning to become certified. I am pleased with my choice of instructor. Her strengths are my weaknesses and we complement each other. I make good progress and never mention my riding lessons at work.

Life can change in the blink of an eye. On a pleasant July morning, as I pull Scottie up and start walking home, I hear someone calling "loose horse." The horse in question has dumped his rider and is quietly trotting home. His exercise boy races down the track past me, ducks under the rail and manages to cut in front of his horse. By waving his arms and yelling, he succeeds in turning his horse around, straight back into Scottie.

Scottie panics as this horse races towards us. My horse wheels towards the outside rail of the training track. For a moment, it feels as if Scottie is going to jump but he unceremoniously dumps me over the rail. At this moment, the outrider shows up. While I easily catch Scottie, I turn down his offer of a leg up knowing I will only get dropped again. Rather embarrassed, I lead Scottie home.

Now Scottie is not only afraid of horses galloping up behind us, he is terrified of horses coming at him.

Since we are close to a race with Scottie, Red wants to school Scottie in the paddock one morning. I gallop him around to the front side of the main track and pull up. I am met by Red and Debbie, Scottie's new groom. Debbie clips a shank onto the bit and leads us into the saddling enclosure. Red is pleased at how well the horse is behaving.

"Put him in one of the stalls," he directs us.

Debbie leads me into one of the saddling stalls.

"Turn him around. Good. Now, Jan, jump off."

I hop off the two year old. "Okay, Debbie, throw her back up."

"Debbie, turn him to face back in," I motion to the groom.

"Whaddaya mean, turn him around?" Red jumps on our case.

"We always mount facing a wall. Otherwise, he'll buck me off," I reply defensively.

"He's not going to buck you off." Red is adamant.

Uh-oh. Red is in one of those moods. Debbie and I exchange glances. We know enough to shut up and follow orders.

Debbie boosts me up. As I land in the saddle, Scottie starts bucking. On the third buck, Scottie hits the end of the shank and Debbie pulls his head around. With no chance to tie on, my body continues in a straight trajectory and we part company. I dust myself off and bite my tongue to prevent myself saying, "I told you so."

"Debbie, face him in. Give Jan another leg up," Red harrumps. This time I stay on. I have the satisfaction of knowing this will prevent the jockeys from being thrown off in the future.

In short order, Scottie's Jockey Club papers are returned with his official race name (Rapid Fire), he is tattooed, okayed from the gate and his workouts are recorded. He is entered in his first race, breaking his maiden first time out and winning by a nose.

Next morning Scottie is hand walked for half an hour. The following morning, Red asks me to take him out to the back field and ride him once around. The back field is now my favourite place to ride. Most horses also love the change of scenery. So just after six a.m. I stroll down the main road toward the field. I notice a horse and rider exit the barn on my right, heading for the training track. Scottie sees them too and tenses up. As they turn the corner toward me, Scottie, anticipating being run into again, freaks out and rears straight up. In the blink of an eye, I realize my horse is coming over on top of me and I let go. I slide over his tail and land in a heap.

Scottie twists in mid air to regain his balance. Then, using my right leg as a starting block, he gallops over me and runs off.

A trainer who witnessed my accident runs to assist me. As he helps me to my feet, my right leg buckles under me. As I look down, a goose egg is already forming above my knee. As I sink to the ground, I ask the trainer to call the ambulance. There is no sign of Scottie. I pray he does not injure himself. This horse does not have much of a self preservation instinct but there is nothing I can do.

I hear the ambulance paged to my location. A couple of minutes later it arrives. The two professionals help me into the back of the vehicle. These guys may be trained to deal with the critically injured but they sure lack some first aid knowledge. One of them rolls up my jeans until the fabric painfully constricts the swelling like a tourniquet. While my instinct is to kick the attendant in the teeth, I behave and ask for a blanket to drape around me. I drop my jeans. The swelling is getting pretty impressive. I suggest some ice might be in order. They have no idea where the ice is. I point out the ice hut some thirty feet away. The driver walks over and comes back with a twenty pound bag of ice. We dig some ice out and find a bandage and I make an ice pack. Seeing that I am not critically injured, the attendants drop me at my barn, leaving me to arrange getting myself to the hospital.

I am relieved to learn now that Scottie has already been caught and returned uninjured to the barn. A groom drives me to the hospital ten minutes away. It is so early that I have to wait for a technician to start work before anyone can take an Xray. My watermelon sized thigh now has a perfect horse shoe imprint in it, filled with a bubble of fluid. The leg is not broken; just badly bruised. I am sent home with crutches and pain killers.

I spend two days on crutches and another ten days at home, laid up. I have not had this much time off in years. My injury forces me to slow down and to think seriously about the state of my life. I am no longer satisfied to settle to be second best and set new goals for myself. The guy I am currently seeing never bothers to come by to see how I am so I dump him. I decide that as soon as possible, I will resume my riding lessons and try to get certified as an instructor. I will do what it takes to get out of this rut.

Coincidentally, Shelley offers me an apartment at her farm. Much as I love living in this apartment, I need to make big changes. I go to see the apartment and decide to rent it. Sandy is stunned when I break the news that I am moving out. We give our notice to the apartment management.

Once I can walk again, I help out in our barn by walking hots for a couple of weeks. I am still unable to gallop because of the location of the bruise.

Scottie has run again, finishing second and was claimed so I never have a chance to ride him again. I don't know where the horse ends up but I will carry the lump on my leg from him to my grave. I could kill the idiot rider who caused all this.

When I get back to riding lessons, my instructor Sue tells me the date of the evaluation for the instructor certification is set just a month away. Do I want to take the evaluation exam? Do I? You bet! Sue books me in for extra lessons and I set to work prepping for the evaluation. I work harder now than I ever have in my life. I am galloping in the morning, riding a horse on the farm for Shelley in the afternoons, practise teaching in the evenings with my instructor and doing extra lessons to brush up for my exam.

It is late in August now and we are low on horses to gallop with most of them racing weekly. Only a couple of weeks after I return to work at Red's, he lays me off with a week's notice.

While this will give me extra time to prepare for my instructor's exam, there is still work to be finished in his barn. New Gold, the goofy gelding that I gallop, is due to race the day after I am finished. We've made an observation about Newfie. Although he had no real speed and can only work a quarter mile in about twenty four seconds, which isn't particularly fast, he can keep up the same pace seemingly indefinitely. Theoretically, if the race is long enough, he will win. So when we find a race at a mile and three sixteenths on the grass, it is as if the race is written for him. Now, with the race only two days away and nothing to lose, I approach the boss to confess my earlier error.

"Red, do you remember you were in the States when Newfie won? And that I was supposed to gallop him along?"

"Yes."

"Well, I forgot. I just galloped him two easy miles."

Red ruminates for a moment. "Well, Jan, I'll tell you what we'll do," he says conspiratorially in his raspy voice, "You gallop him tomorrow and let him two minute lick for five eights in the second mile." He is asking me to increase the pace on the second lap of the track.

Next morning, the groom leads us out to the training track. Newfie is so fit now he doesn't want to settle into an easy gallop. He keeps trying to take hold of me. While I know I can slow him down, I am sure he will only sulk when I ask him to gallop along and not really put out for me. So I 'play him like a fish' as Jackie would say, easing him out, reeling him back, softly rebalancing him, keeping him quiet. We canter like this for a full mile and a half. Then, as we approach the five eights pole the second time, I reach for a shorter hold on the reins, fold low onto his mane and let him gallop along.

He changes gears. The rhythmic canter morphs into a piston-like gallop As the pace picks up, my vision narrows to a tunnel before us and everything to the side of us becomes a blur. There is no feeling on earth that can compare to a galloping horse. It is like a jet leaving the ground, only better. This is what racing is all about. Newfie rates beautifully. Not full out, as in a workout but leaving plenty in reserve.

As I pull up, I am thrilled with him. I turn for home, standing in my stirrups and balancing on the breast strap and let him jog. Newfie has no intention of behaving now. He is so full of himself, bucking and kicking out that it is all I can do to stay on him. Just a couple of weeks ago, he managed to drop me by shaking himself like a snake. I don't want to lose his race now.

Red meets me at the stall as I unsaddle. "How'd he go?" he demands.

"Great, Red. He's really ready." I'm beaming.

"I thought you went faster the first mile than you did the second," he growls.

"No, Red. I could have rated him more but then he wouldn't have run when I asked. He'll be tough tomorrow."

Red is still fuming as he stomps off. However, this is my last day at work. I don't have to worry about Red's moods now.

It is several days before I lay my hands on a racing form and read the results. It puts a smile on my face.

At the first quarter, Newfie is on top by three lengths. Still on top by three at the half mile. Ditto at the three quarter pole. I can imagine the other jockeys thinking, "He'll come back" and let my golden horse set the pace.

At the mile mark, his lead increases to five lengths. The other jockeys are still waiting for Newfie to tire. He wins by thirteen lengths. If I know Red, he had a pretty substantial bet on our horse. No doubt Red laughed all the way to the bank.

I do not waste any time hanging about the track. For the next six weeks, I work harder than I ever have to prepare for my instructor's exam. I do extra riding, more lessons, practise teaching and bone up on my stable management. The exam is nerve wracking. The wait for results even more so. Finally I learn that yes, I have passed my riding instructor's exam.

Chapter Thirty-One

NEVER RIGHT

"Hey, Jan, how'd you like to come work for me as an assistant trainer," Shelley yells up the stairs to my apartment. "Wow," I think to myself. "What an opportunity! This is too good an offer to pass up." I accept the offer immediately.

With Woodbine reopening for training, I give my notice at the dressage barn where I have been grooming all winter. "I knew it! The good ones always quit," Carol, the coach, cries. She is upset to lose me but wishes me well.

When Shelley ships her horses in, I start back with her. Our stable has more racehorses in training with her this year. Shelley hires Wanda, an older woman to groom and also hires an exercise boy.

From that moment on, I can do no right. Far from being the fun and fair boss that I worked for before, Shelley picks on one or another of us. If I am five minutes late, the earth caves in. If I leave a piece of straw in a tail, I am incompetent as a groom. I can't bath a horse properly or turn one out for galloping to suit her. Since I have always been taught not to overfuss the racehorses, I find this baffling. In fact, I can hardly believe this is the same person I worked so happily with. I am sure nothing would make Shelley happier than if we all fought among ourselves.

I start dating a trainer whose horses are in the next aisle. I am only moderately interested in this man but I find his attention flattering, a balm against the stresses of work. I keep our liaison a secret, never divulging to Shelley where I am going in the evenings after work. Shelley is an eager gossip and I don't like my private life splashed about the track.

With the advent of twilight and night racing several years earlier, it resulted in much longer work hours and many experienced grooms have left the track. The influx of young girls eager to learn about horses has also seriously declined. The result is that help is now hard to come by. Shelley hires Wanda's daughter to groom even though the girl works afternoons at the races. While Wanda is sweet and I have no trouble getting along with her, her daughter, one year younger than me, is a spoiled brat. We had stall guards (rubber barricades similar to webbings) in pink or blue. Within days, the girl has 'traded' me for all the pink stall guards and taken them for herself. I just find her incredibly childish.

We ship into Greenwood, where we are adopted by an abandoned cat. He holds up a paw that has a split in his pad and the four women in our barn immediately fall for his ploy. Gimpy, as we name the cat, immediately proves his worth because Greenwood is overrun with mice.

The trainer I am dating stays back at Woodbine but I continue to drive to his place several nights a week. I enjoy his attention and it provides some balance from work for me.

Wanda and her daughter both work at the races so all of the afternoon feeding falls to me. In fact, I am now getting on a couple of horses, rubbing four and walking three every morning plus all the feedings. In addition, I run all the horses in the barn. I am doing work that is the equivalent of two and a half full time people. Still, nothing I do pleases Shelley.

One of the horses I rub is a huge 17.3 hand gelding named Farriers Gold. When he races one afternoon, he comes out of the gate crooked. Farrier jumps to the right before straightening out and running with the pack to finish second. When I pick him up after the race, I see his foot bleeding a little.

Farrier has 'corked' himself with a back foot coming out of the gate. Race shoes can have **caulks** in them for traction. The idea is similar to cleats on golf shoes. He dug one of them into his **coronary band** when he jumped sideways.

I walk the gelding back to our barn where Shelley helps me bath him. Farrier keeps holding the injured foot as high up in the air as he can. Shelley sets up his drinking water and I begin cooling him out.

"Would you like me to do him up in a **Furacin sweat** and give him some **Bute**?" I ask.

"No," she replies. Just do it up in a Furacin wrap."

"Are you sure? He seems pretty sore to me."

Shelley is adamant. She soon leaves to drive home and leaves me to cool out the horse.

The problem is, Farrier will not cool out. Every time we stop at the water bucket for a drink, he lifts his sore leg up as high as possible. His body is still hot to the touch after nearly an hour. I decide to defy Shelley. I give the horse a second bath to cool him as best as possible. I dig out the Bute and mix it in his grain. I do the leg up in a Furacin sweat.

For the next two days, the leg is fine and I hand walk the tall horse for his exercise. On the third morning, however, Shelley pulls his wraps off before I get to the barn. She sees that I have disobeyed her and she is furious at me for ignoring her. Now I have no choice but to follow her orders and wrap the leg dry.

Next morning, three days after the race, the leg blows up over night. It is hot, tender and swollen. I can't imagine what the leg would have looked like if I had not treated it initially. Shelley calls the vet in to see it. The vet examines the leg and prescribes – you guessed it – Bute and a Furacin sweat. Shelley never apologizes for being wrong. I have known for quite a while that I know more about keeping horses sound than she does. It is not my place to say so. I begin a slow burn.

We finally get the horse sound a few days later. The horse has already had a couple of extra days off. Instead of galloping him, Shelley asks me to jog him two miles the wrong way of the track.

He is already fit enough and she just wants to be careful with his injury. I trot him once around. As I approach the out gap, out of the corner of my eye, I notice a half dozen people sitting on the outside rail. Feeling fresh, Farrier launches into a bucking spree. As if choreographed, six people stand as one and step backwards over the rail.

When a horse is 17.3 hands high and wants to put his head down and buck, he can do just that. He loosens me on the second buck but it isn't until the sixth jump that I am sent *backwards* over his tail into the dirt. He is the only horse I have ever fallen off backwards.

Credit: *Michael Burns*

When a horse is 17.3 hands high and wants to put his head down and buck, he can do just that. Danny Beckon up.

Of course, Shelley is furious at me for coming off. She is more concerned about the horse being injured than about me. I never let on that I have hurt my back. By the next day, I cannot lift my arms above my shoulders but I am not about to let her know. I still have the strength to gallop; those muscles are unaffected. I find a chiropractor in the area and get myself in for a treatment.

The chiropractor is horrified at the amount of Xrays I have had taken in my lifetime. After bawling me out, she insists I dig up Xrays already taken of my back before treating me. It is a couple of days before I get relief for my pain.

I am still working six days a week, galloping a couple horses, rubbing three and hot walking several every morning and feeding the entire barn every afternoon, plus running all the horses.

Yet Shelley continues to cater to the whims of Wanda's daughter. At the same time, Shelley continues to state "Nothing ruins good help like bad help." I do more than twice the work of this girl, yet Shelley is afraid that the brat will quit?

I am struggling to find the courage to stand up for myself. Letting people walk all over me is such an old pattern and I am determined to free myself. I also need to break up with the trainer I have been seeing. I see too many similarities in behaviour between him and the Boyfriend. If I get more involved with this man, I will be subject to the same manipulations I suffered through before. I am not quite ready to leave him yet; his support will be invaluable when I quit Shelley. I decide to drop by the trainer's barn on the way home. I am feeling pretty low and can use some cheering up.

When I get to his barn, I am greeted by the concerned groom asking, "Have you seen him?"

The trainer has not been to the track for two days nor has anyone heard from him. Phone calls have gone unanswered. I have to run a horse early today and can't run up to the trainer's farm now. The groom decides to go for himself.

Late that afternoon, on a beautiful spring day, I drive up to the farm as well. As I travel down the long driveway, I am met by the groom some distance from the house. He stops me to tell me the trainer has committed suicide and prevents me from driving closer. The police are investigating and there is no need for me become involved.

I am stunned. The violence of the suicide contrasts sharply with the peacefulness of the warm afternoon. Birds are singing. The grass is coming green. I am unable to wrap my head around what has happened. While I planned to leave him, I still harboured some warm feelings for the man. I drive over to a friend's house to cry on her supportive shoulder. I need someone to help me with this shattering event.

With all the difficulties I am having at work, I never divulge this death to Shelley. I am too distanced from her to open up my private life. Only my closest friends are privy to my confusion now.

At the funeral, I run into the man's ex-wife, a girl I know slightly. We slip away from the crowd to have a heart-to-heart talk. As she relays the difficulties she had with her marriage, it confirms to me that this was not a person I wanted to be with long term. While I am sad for the ways things work out, I am also relieved to be out of this affair when his ex-wife shares how he stalked her. These opposing emotions only deepen my confusion.

Now is not the time to make a decision, any decision. For a couple of weeks, I am glad of the structure that work provides. I grit my teeth and hang in a little longer. Shelley is unaware of my demons, even though I still live on her farm. I have learned well how to cover my tracks.

The strain of trying to do so much work finally becomes too much for me. One afternoon, as I am feeding, I break down in tears from sheer exhaustion. The stresses of the past few months are too much for me. I am tired of being used and abused. I want to be respected for all the knowledge I have built up.

I do what I should have done a long time ago. I quit.

Chapter Thirty-Two

Track Life

Although Woodbine is located on the edge of a major metropolis of a couple million people, the backstretch is incredibly isolated from daily life. Cell phones do not exist yet. There isn't a regular transit to connect us to the outside world. World events rarely intrude upon us nor are any daily newspapers sold on the backstretch. All that matters to us is racing and the horses themselves.

This isolation has created a community where the veneer of society never quite adhered. We swear at those who irritates us then promptly forget our anger. We take our behavioural cues from the horses, living in the moment with no thought for our future. Every day is the same as yesterday and time is defined only by the seasons.

There is a well defined hierarchy to the backstretch yet everyone acknowledges one another with a nod or a friendly 'hello'. Today's lowly hotwalker may be next year's hot jockey. That old man who owns a single racehorse may have trained a Queen's Plate winner. We are all part of this soup of the backstretch, where even your name may be unimportant. I, like most of the increasing number of women working at the track, have adapted to this lifestyle so completely that I rarely question it. I only notice the oddness of our collective behaviour when something occurs to call attention to it.

I am still at work when a stranger appears in our barn, asking for 'Leslie'.

All of us look at each other in surprise. "There is no Leslie working in this barn," we reply.

"Cuzzy Bear, then," is the comeback.

"Oh, Cuzzy Bear, you'll find him in his tack room, third one over on this side."

Cuzzy Bear is one of those oddities common around the track. I am not sure he is right in the head – I have never overheard a whole intelligible sentence from him, let alone an actual conversation. This old man walks hots, mumbling incessantly to himself, "Bo-wo. Bo-wo. Cuzzy bear. Cuzzy bear." He never yells, "Heads up," when taking a short turn, forcing me to listen for his monotone chant to avoid him.

Cuzzy Bear is little different than many of the older men I know on the track. The racetrack is his whole world. He works and lives here, doesn't own a car, may even be illiterate. I have never even thought about him having a Christian name until now. I'll bet only half a dozen people on the track know his real name, including the trainer who signs his pay cheque.

This is typical at the track. Unlike sports figures who acquire their monikers from the writers who follow their exploits, I only know these racetrackers by their acquired nicknames – Jake the Rake, Cuzzy Bear, Duck, Fourfoot, Peter Pig. I never question how these people lost their identities but they are the fabric of the backstretch itself.

The backstretch, on any particular morning, is a hodgepodge of society's strata. While most people find their own compatible crowd soon after leaving school, the backstretch brings an odd assortment of people into close proximity. Here, people are not judged on the clothes that they bedeck themselves with. The rich owners believe they sit at the top of the hierarchy of the backstretch but in actual fact, they occupy the lowest rung. They are kept out of the loop of what actually occurs in the barn. What really matters to horsemen is your skill level. If you apply for a position, the trainer asks if you can do the job, then immediately puts you to work. Some grooms have not even completed public school.

I pull into a barn one day, hop out of my car and absentmindedly lock the door with the engine still running. When I discover what I have done, I do the only thing possible. I walk back into the shedrow and bellow, "Can someone break into my car for me? A large portion of the backstretch population has a criminal record and the likelihood of someone in the shedrow having the skills to open my car is high. It's better to be friends with everyone around here than to have many enemies. Sure enough, within thirty seconds of my call for help, I am already driving away.

Lack of social skills is no impediment to climbing up the ladder of success. Jimmy, the trainer who is stabled beside us, may be the most obtuse person I've ever met. My efforts to befriend this man are met with indifference. Taking a cue from others who drop by, I try greeting him with insults instead. Jimmy promptly warms to me and becomes friendly.

One day Jimmy relays this story to me. He buys a pony for his five year old daughter. When Grandma and Grandpa come to visit, the child proudly leads her forebears to see her pony. When the pony proves to be a bit naughty, the diminutive child hauls off and kicks the pony in the stomach, yelling, "Stand up there, you goddamn sonofabitch!" While Grandma turns blue and keels over, the child explains, "Well, that's what Daddy does."

Many of the kids that come to the track do not initially fit in. They are hired straight out of school and have never had to think for themselves. With these naive city kids, it is common to teach them to question orders by sending them on wild goose chases. Most of the younger grooms willingly participate in such silliness.

I am just finishing up in my barn one morning when I hear a groom across the way say, "Oh, she'll have one," pointing to me.

I catch on immediately. "What's he looking for?"

"A saddle stretcher."

I wave the naive hotwalker to follow me and march down the aisle to the tack room. I open our biggest trunk and begin rummaging around the bottom of it where the kid can't see me laughing. Finally, I straighten up, wipe the grin from my face and pronounce, "We must have sent it back to the farm."

"Funny, you're the fourth person I've asked," replies the hapless hotwalker. I send the clueless kid down the aisle to another barn.

There is no such thing as a saddle stretcher. Nor bit injectors, left handed pitchforks, post holes, etc. After spending a couple of hours trudging around the track looking for a mythical item, most kids will finally realize the joke is on them and become a bit more streetwise.

A new hotwalker is assigned to cool out a horse in the test barn. The kid dutifully allows the horse to takes sips of water from the assigned pail until the horse is watered off. The attendant directs the hotwalker to turn the horse into a stall. Once the urine sample is obtained, the hotwalker asks, "Now what?"

"Now you go home," the attendant answers. So the hotwalker does. After a couple of hours, the groom comes looking for the animal, wondering why the horse never returned to its barn. She finds the animal still in a stall in the test barn. The hotwalker left the track ages ago.

The kind of problems we experience at the track are completely different from what most people do to earn a living. I can no more grasp how to survive in a poisonous office environment dressed in heels and fashionable clothes than most people would warm to my life, outside in all weather, struggling with large and sometimes difficult animals. I have spent almost all my career working in non-traditional work. The following incident illustrates what I mean.

It's mid March and I am standing in line in the Racing Commission office, waiting to get my photo I.D. processed for this year. Behind me, I can hear a young trainer talking to his buddy. The trainer was recently stabled in the eastern United States. One chilly afternoon, he shipped an old claiming horse into a second rate track for a race.

The trainer and his groom put the horse in a stall in the receiving barn. They filled a tub of ice and water for the horse to stand in, covered him with a cooler and they buggered off to the warmth of the kitchen.

It was a raw spring day with a chilly wind making it feel even colder. The two men sat in the kitchen, nursing their coffee, not paying attention to the time. When the call came to get ready for their race, they were caught unawares. They hustled back to their barn only to find the ice had frozen into a solid block around the horse's legs!

By the time they had chipped the ice off the legs, the race had already run. The trainer was fined two hundred dollars for missing the race. (I had never heard of the ice freezing onto a horse's leg. No vet can tell me why the poor horse never gets frost bite, either.)

These little vignettes flavour my working days. The backstretch is unlike anywhere else I have been. Life here is raw, unfiltered and often quite humorous. Just how I like it.

Chapter Thirty-Three

WHAT A TRIP

When I quit Shelley in June, I decide to take a holiday. I haven't had time off in a dozen years, except for the time Scottie stepped on me.

On the phone to Cindy, she asks, "Would you like company? I can book my holidays anytime I wish."

"Sure, I'd love company."

"Have you got any idea where you would like to go?"

"Yes," I confess. "I want to go to Kentucky and see Secretariat."

This is a bold statement from me. Secretariat, winner of the Triple Crown in 1973, is the most famous racehorse in America. The big thoroughbred farms do not let the public just wander about their farms and have access to their valuable stallions. Yet, I am hoping to pull off a miracle.

We discuss our plans. Cindy books her time off. We will take her car and decide to camp.

I have no savings to draw upon. I only have one hundred dollars in my wallet and no credit cards of my own. I cross my fingers this will be enough money.

We drive south to Kentucky and we camp near the Kentucky Horse Park. We begin our day at the brand new park as the most likely place to get information. Besides, I have not seen it yet. At the admission booth, we enquire about tours of the thoroughbred farms.

"Most of the farms no longer allow visitors but Spendthrift Farms is a public stable and has three tours every morning. The last tour today leaves in fifteen minutes." The farm is five miles away.

I look at my watch. "We'll make it," I declare. We grab a map, drive like maniacs and just squeak onto the tour as it is starting.

The Triple Crown is the pinnacle of thoroughbred racing. The three races are on different tracks in a five week period open only to three year olds. Eight horses had won the Triple Crown up until 1948, then there was a twenty five year drought until Secretariat made the record books in 1973. Two more horses, Seattle Slew and Affirmed, also completed the feat in the next five years. Now all three stallions have retired to stud in the Versailles, Kentucky area. Spendthrift Farm is the home of Affirmed.

Spendthrift has always been one of the big names in breeding farms. Many great horses have been bred there. Our tour begins with a film about the farm and its stallions while seated in the plush theatre of the sales pavilion. Then we board a shuttle to ride round the lovely farm. The highlight is seeing Affirmed, winner of the 1978 Triple Crown. I watched the famous duels between Affirmed and Alydar on television at Bobby Fisher's my first summer at Woodbine.

Affirmed proves to be a surprisingly feminine chestnut stallion. There is nothing of the beefy and proud stallion I am expecting. He is slightly built and seems very quiet. After meeting Affirmed, I am not surprised that Alydar proves to be the better stallion. I snap a photo.

At the end of our tour, we question our tour guide. "Do you know of any other farms that offer tours?"

Unfortunately, she does not but as she is talking with us, another tour guide overhears the conversation.

"I do private tours," Joan states and hands us her business card. I recognize her last name as being associated with some of the well known racing families. Her rate is twenty five dollars plus a five dollar tip to each stud groom who shows us the stallions. We happily agree to go with her next morning. What do we want to see?

"Secretariat, at Claiborne," I state emphatically.

Joan says she can get us in there tomorrow morning. I am over the moon.

Cindy and I talk over our next move. She has already seen the Kentucky Horse Park. We decide instead to drive three hours across the state to see the famous Mammoth caves.

That evening, Cindy and I pool our change to phone Joan and firm up our plans. We agree to meet by nine in the morning. Cindy and I now drive back the three hours to the Versailles area. We don't want to be late for this important meeting.

It is midnight when we pull into our campsite and quickly raise the tent. We are soon asleep.

Joan is as good as her word. She is getting us into Claiborne and another farm I have never heard of, Three Chimneys. She offers us the choice between Keeneland Race Course or Castleton Farm, where the legendary Standardbred Bret Hanover is standing. Bret Hanover is as famous to Standardbreds people as Secretariat is to Thoroughbreds folks.

Cindy and I are so very different from Joan's other well-heeled clients. For one thing, we have very little money. Also, we have both worked with Thoroughbreds and are familiar with the bloodlines and the actual animals themselves. Joan finds us such a refreshing change that she takes a shine to us.

I have been to Kentucky before. My family vacationed there when I was a teenager. Now the miles of picturesque white fences have been replaced by black creosote ones, to discourage horses from chewing the fences apart. The perimeters of the farms are still ringed with handsome low stone walls. Joan explains that originally the walls were much taller. The limestone underneath the state of Kentucky builds the strong bodies of the horses but is easily leached by water to form caverns and pockets underneath the state. The walls have sunk into the ground over the century since they were built.

As we drive down the narrow back roads, Joan informs us about the farms we are passing.

"Do you see the weather vanes on the barns on this farm? They are gold plated and cost thirty thousands dollars each. The farm is owned by an Arabian sheik, the seventh richest man in the world. When he flies in to the yearling sales in his private jet, he pays cash. Imagine that. Paying for a million dollar yearling in cash. He just opens a brief case and pulls out the money."

We pull into Claiborne Farms, the most prestigious breeding farm in the world. While we wait outside the stud barn, the stallion man puts a shank on The Big Red Horse and brings him outside.

Secretariat is a ham. He immediately poses for us. Obviously, he is used to being shown off. Then he gently paws the ground with his foreleg, begging for peppermints.

"Would you like to have your photo taken with him?" Joan asks us.

Cindy and I look at each other in disbelief. Does it get any better than this?

The well behaved stallion stands quietly as we position ourselves in front of him. Joan obligingly presses the shutter button. We tip the stallion man and thank him profusely.

Joan whisks us off to our next stop. Three Chimneys is a new farm and is home to the third living Triple Crown winner Seattle Slew. We begin our tour in the handsomely appointed office/visitor centre. The staff explain how, when Queen Elizabeth visited last year, the staff picked the dandelions from the lawn on their hands and knees.

Would you like to have your photo taken with him? Does it get any better than this?

What a Trip

We are invited to sign the guest book. We are being treated like royalty ourselves. For two little nobodies from Canada, this is over the top. This small farm only has a single barn for the stallions. The barn is stunningly beautiful with a beamed cathedral ceiling. The mares are shipped in each day to be bred and vanned back to their resident farms the same day. The barn is empty because the stallions have already been turned out. At this farm, all of the stallions retired sound. They are ridden for exercise each morning to keep them breeding fit (imagine riding a Triple Crown winner as part of your chores!) We walk down the path between paddocks to view the stallions. The fields are brown due to a prolonged drought, which has the native Kentuckians worried. The stallion man talks about the farm and its horses.

Lastly, he proudly shows us Seattle Slew.

The heyday of Thoroughbred racing in America was the 1930's, when racing was the most popular sport in the country. But the 1970's gave us a second brush with greatness, with three stallions managing the almost impossible feat of winning the Triple Crown within the decade. Within twenty four hours, I am able to view all three of them. With the last of my film, I snap a shot of Seattle Slew in the field.

I don't even have enough money left to buy another roll of film. Joan is so charmed by Cindy and I that she decides to take us to both to Keeneland Race Track and the Castleton Farm.

Castleton has an old world aura about it. Many of these Kentucky horse barns date back decades. There is already a life sized statue of Bret Hanover erected on the lawn, which will grace his grave one day. As we drive through the farm, Joan points out the two newly restored slave quarters. They are a window back in time. The farm took the derelict structures apart, numbering each board as they went and rebuilt them exactly. The slave quarters are tiny and I wonder how many people were crammed into them. We see the famous stallion Bret Hanover and continue the tour.

Keeneland Race Track is a time capsule from the 1930's. It doesn't even have a P.A. system. No wonder it is the choice of film makers when making period movies about racing. It is such an intimate space and I love it.

Joan drops us back to our car. We thank her profusely as we pay her for her time. She has made our dreams come true. Over lunch, Cindy and I decide to pack up and head home. Within a couple of hours, the longed for rain comes. And comes. And comes.

By the time we are driving through the flat lands of Ohio, the water has filled the ditch between our lanes and the oncoming traffic and is being whipped into waves. On the right side of the road, the page wire fencing is half buried in water. Traffic comes to a full stop. For half an hour the cars do not move. Finally we begin to creep forward.

We come to the cause of the delay. The road is flooded about two feet deep. We have to carefully drive a couple hundred feet through the water and then be diverted onto the exit to go around more flooding. Only the transport trucks are being allowed to continue on the highway past the exit.

Cindy's little Honda is a low slung car. If the water splashes up onto the motor, the car will stall out. We would have to be towed out and wait for the engine to dry out. This would mean renting a hotel. Cindy, at least, has a credit card she can use. I have only a few dollars left until we get home. We say our prayers. The gods are still smiling on us because we make it through. Although delayed, we are able to cross the border that night.

It has been an amazing journey and remains the highlight of my life.

Chapter Thirty-Four

WE, THE ANONYMOUS

There is a mystical side of my personality. For example, through journaling, I've come to the conclusion that my moodiness is often a reaction to big changes in my life ahead of the actual events themselves. I am curious about what pops up in the ramblings of my diary. I record my emotions, hopes, dreams from the night and my musings about them.

Lately, I been having a series of dreams in which I riding on a train. *When the train reaches end of the line, I quietly disembark.* I am puzzled by the reoccurring nature of this dream, which does not feel threatening in any way. Eventually, I see this dream indicates to me that my learning on the track is complete. My soul is calling me to new adventures. I just don't know what they are yet.

Shortly after quitting Shelley, I am driving down the road past Red Millar's barn when the old trainer hails me.

"Hey! What are you doing for the next couple of weeks?" he hollers at me.

I pull my car over to his barn. "Not much," I answer.

"How'd you like to gallop for me for a couple of weeks. My [exercise] girl is getting married, then going on her honeymoon."

I agree to help Red out. I have been wresting with what to do with the rest of my life for a couple of weeks now. This allows me to drop back into the comfortable routine. While it's lovely to be galloping again, most of my former co-workers from this barn have moved on. One groom has retired, another married and left the track while the third groom has moved to another barn. I am an outsider now with no real connection to these people. In fact, the racing industry no longer serves it purpose for me. I have learned about all I can about horses here. These insights reinforce my decision to move on as well.

I will never regret the time I spent in the racing industry. My ten years here have changed me profoundly. Woodbine has gifted me with enduring friendships and healed me from deep scars on my psyche. I have grown into a skilled horseman. I take with me a rich bank of memories – frosty morning gallops; breaking from the gate; paddocking racehorses on sultry afternoons. A letter my sister once wrote to me sums up my feelings about the time spent here.

I think about you lots. It must be tough for you to be working in the field that you want to and not be able to have the acquisitions that are the "marks of success' in our "success oriented family"...There are no perfect choices. Choices always involve sacrifice and everyone struggles with these issues again and again at different times All involve giving up something. While I don't have absolute control over what happens in my life, what twists and turns there are, I do have tremendous power of choice....Taking risks and following one's heart just aren't high priorities with (Mom and Dad) and really I don't think they ever were...So it's hard for most of the family to understand your putting other things ahead of financial security and affluence. The acquisition of knowledge, skills, success in relationships, a sense of accomplishment – these are the things I value most and of course, having good friends. If I had to trade these for money, I would feel so empty. And I know there are experiences for you that make you feel rich. Would a mansion with a pool take the place of those early morning waking up activities at the track, warm and familiar greetings of friends, the feel of a good colt underneath you, the pleasure of successfully breaking a young horse, etc, etc? Dreams never evolve exactly the way we dream them. But precisely therein lies the thrill in realizing them in fits and starts. When some momentary experience is "just right"' it is worth all the struggle and setbacks it took to realize the dream.

I am so very different from the person who first walked through the gates of the backstretch. Coming to Woodbine was a decision made out of desperation for a job rather than any grand life plan decision. I just wanted to become an expert with horses. At Woodbine, I have been accepted into a community bonded by our common love of horses. I found a lifestyle on the backstretch that is unique, a throwback closer in time to that of the 1940's.

After leaving the track, opportunities to teach riding open up easily for me, now that I am certified. My forner instructor hires me part time and helps me to find other jobs. Other coaching opportunities come to me as well. For the next thirty years, this is how I make my livelihood.

I always planned to write about my experiences of the backstretch, the gated community that few will ever experience. Although I have a few photographs of my own, I was hoping for better pictures of the day-to-day life on the backstretch to flesh out my book. I contact the archivist for the Ontario racing industry Tom Cosgrove and drive three hours to Woodbine to meet with him. When I explain what I am looking for, Tom informs me that the type of pictures I am looking for do not exist at Woodbine. Nevertheless, Tom offers me any assistance possible and leads me through the bowels of the grandstand to the archive room. It is a small space, about the size of my kitchen. The Woodbine Entertainment Group (formerly the Ontario Jockey Club) only keeps files on the famous horses and people plus the results of their racing. Anything else is discarded due to lack of space. While disappointed by the lack of photos, I'm not surprised.

We are the anonymous, the grooms, hotwalkers and exercise riders who prepare the racehorses for training. Many have dedicated their lives to the well being of the horses in their care. This is their story, as well as my own. Yet my story is everyone's story. It is the mythical hero's journey of coming of age, of making mistakes and falling down and getting back up again. I have learned my lesson well. I know now that if I can do the impossible just once, I can probably do it again. It requires digging deep within myself to find the courage to face myself. I have claimed my power and no longer let people's opinion of me limit my world.

My final day at Woodbine arrives. When galloping is done, I clean my tack for the last time and hang it up. Red shakes my hand as he gives me my final cheque. There will be no elaborate goodbyes. Today, after the races are over for the day, when the trainers have packed up and gone home, when the grandstands empty, littered with useless betting stubs; the grooms will still be found squatting in the straw, poulticing the legs of their racehorses. At first light tomorrow, the exercise riders will be getting legged up on their first mount of the day. But I will not be there.

I have already moved on.

Glossary

Agent, jockey's – almost all jockeys are represented by an agent, who books their riders and receives 10% of their earnings. An agent can only represent one apprentice and two journey men.

Also eligible – horses that are entered in an overfilled race that can be used if other horses scratch (withdraw) before the next day.

A meet – the higher end racing. Woodbine, Greenwood or summer at Fort Erie.

Ankles – wind galls or osselets (puffy ankles or bony arthritis).

Appaloosa – the breed of horse characterized by their spotted coat.

Babies – either yearlings or two year olds.

Backing – getting on a horse for the first time.

Bay – brown with black mane, tail and maybe legs.

Belly, to – lying across the saddle, before mounting a youngster for the first time.

Big horse – the horse in the barn that is making the most money, most valuable horse.

Blaze – a wide stripe of white on the face.

Blow – losing (a race).

Blow out – work a horse.

B meet – secondary level racing. Lower purses and lower claiming races.

Bran mash – a mix of grain, bran and hot water to keep a horse from being constipated.

Breaking – properly called backing. Getting a horse to accept a rider and take some direction from them.

Break one's maiden – win its first race.

Bridge – holding the reins crossed at the ends, so there is two reins in each hand (double cross) or just doubled in one hand (more common, single cross).

Bottom – fit.

Bought real estate - fell off.

Bowed tendon – below a horse knee there is no muscle, just tendon. These can be injured by careless bandaging, in which the pressure is not evenly distributed or by an unfit horse racing in deep footing. The tendon develops a swelling. This is a very serious injury, often career ending. Also known as hoops.

Bucked shins – a microfracture in the front of the cannon bones. It is very painful.

Bute – Phenylbutazone, a pain killer and an anti-inflammatory drug. It is not legal to race on but can be used at other times.

Buzzer – a device that sends a jolt of electricity out. They are illegal.

By – sired by (a horse is always out of a mare and by a stallion).

Capriole – the horse leaps up off all four feet and kicks out with their hind feet while in the air.

Cast – a horse that rolls over in a stall without enough room to get his legs under him to get up is said to be cast. Often the horse needs

Glossary

assistance to be moved (rolled or pulled away from the wall). Often, you can hear the horse thrashing against the wall.

Caulks – downward metal extensions on shoes that provide extra grip.

Cavalletti – a small jump consisting of two X cross members and a pole between them.

Chain – all lead shanks on the track have a chain about 30" long at the end with the snap.

Chestnut – a yellow to light brown colour of horse with mane and tail the same colour.

Chicken strap – neck yoke of running martingale.

Chin strap – all bridles on the track are fitted with a loose strap under the bit. Its only function is for leading the horse when bridled if there is a rider in the saddle.

Chute – an extension of the track set into the 'corner' of the oval, where the gate is set back.

Claimers – horses that run for a set price in a race. It is a complicated formula that allows horses to run against equal competition and owners a chance to compete where they can earn back some of their investment.

Cleft – there are three grooves in a horse's foot, two on either side of the frog forming a V and one in the centre of the V. Properly called sulcus.

Clocker – one of the officials who work for the jockey club, recording the official times of the racehorses. A racehorse must meet certain criteria before being allowed to race.

Cloudy test – some drug is found in the urine sample of a race horse but it has not been identified. There is something in it but unknown.

Coffin bone – the bone inside the hoof, shaped like a miniature hoof.

Colours - See Racing silks.

Colic – literally, pain in the abdomen. A catch all term for a variety of ailments, many of them lethal. Colic is the biggest killer of horses and is always dealt with as an emergency.

Come up short - unfit.

Commission, Ontario Racing – the branch of the provincial government that oversees that racing is fairly run.

Condition book – a booklet available to trainers that tells what races are available for their horses.

Conditions – the variables that determine a race. These include distance, sex, age, type of race (claiming, allowance, stake), previous winnings, etc.

Conformation – the structure of a horse's body, functionality.

Cooler – a large wool blanket used to prevent the horse from catching a chill or cooling out too quickly.

Coronary band – the soft tissue at the top of the hoof, from which the hoof grows. It is richly supplied with blood vessels and nerves.

Cribbing – a self rewarding habit in which the horse grasps an object with its teeth, flexes its neck and sucks air, making a peculiar burping noise. It causes excessive wear on the teeth, weight loss, more colics plus being an annoying habit. Other horses can learn by example.

Cross (taking a) – racehorse reins are extra long and have a rubber coating for part of the reins. The extra length is doubled. You can hold both doubled sections (double cross) or just a doubled section in one hand and a single in the other (single cross).

Crow hopping – a series of gentle bucks.

Cut a horse – castrate.

Cut irregularly – half the label goes with the sealed cup with no identification on it, the other half is kept back. The groom witnesses this. This is to prevent tampering with the sample. It is then tested for drugs.

Dark bay – black except for brown flanks, nose, etc.

Glossary

Dark day – a day of no racing. At Woodbine, Mondays (except holidays) and Tuesdays.

Deep litter bed – removing only the manure and placing clean straw on top of the old straw. Commonly done in old drafty barns when horses were generally used for work.

Drawing – removing a horse's hay before a race, generally for 24 hours.

Draw lines – an extra set of reins, fastened at the girth, passing through the bit rings and back to the rider's hands. These double your pull on the horse.

Draw, the – a lottery to determine which horses will race and in what position.

Entry – two or more horses in a race with the same owner or the same trainer.

Exercise rider – the person who gallops the horses in the morning.

Flake – hay is pressed into small sections. There are five to eight flakes in a bale.

Float – horses teeth erupt (grow) their entire lifetime. Floating is rasping off the sharp edges that develop.

Foot locker – a small covered box on legs in which are stored the brushes, saddle pads and polo wraps for each groom are stored. It is placed in the aisle opposite the groom's stalls.

Form, Daily Racing – the news sheet that contains the entries, along with their past performances that is used for betting and also for claiming.

Founder - a painful infammation of the feet causing permanent damage and lameness.

Frog – the tough, rubbery V shaped pad in the middle of a horse's foot which acts as a cushion in the foot.

Furacin – a yellow paste with anti-microbial action, widely used on the track.

Galloping – cantering.

Galloping along – a faster pace, an extended canter.

Gap – the breaks in the fence where horses enter or leave the track. Woodbine's training track had two exits, at the 7/8 pole and the 3/8 pole.

Get – all the horses produced by a stallion or mare.

German martingale – an extra set of lines that go from the neck yoke of the martingale to fixed D rings on the reins. These allow you to double your leverage.

Giving a horse a race – using the race to condition the horse, knowing the horse is still not fit enough.

Good for a cheque – in the money.

Greasy – the track condition in which the top is wet and slippery but underneath is firm.

Green - not broke, unschooled, untutored.

Ground driving – teaching a young horse to steer by fastening two long lines to the bit, running them through the stirrups and back to your hands. By walking behind the animal, you can influence him to turn right and left as well as walk forward. A common technique used in breaking Thoroughbreds.

Hack – a ride across country or out on the roads.

Half sister or brother – a sibling out of the same mare.

Halter – a leather apparatus placed on the horse's head for tying up and leading with. Also called a head stall, his hat.

Ham sandwich, not get a – local expression meaning not place in a race.

Hand – four inches (about 10 cm). A measurement of height. Horses are measured at the withers, the high bony spot at the base of the neck. The average thoroughbred measures about 16 hands.

Hay net – a nylon string bag used to keep the hay off the floor.

Glossary

Head number – a small plastic placard clipped to the bridle, corresponding to its number in the program.

Headshy – does not like his ears touched.

Hematoma – a fluid filled swelling, bruising.

Hit the board – finish in the top three in a race.

Hoof pick – a metal tool for removing manure from the feet, used daily. A loop with a hook on it.

Hole, fourth – post position number four.

Hotwalker – a person who cools out the hot horse.

In jail – a claimed horse which must be run for 25% higher for one month.

Inbred – parentage which has close relations. This increases the chances of defects being passed along, such as nervousness, poor conformation.

In gap - the entrance to the track for galloping. There are no gates on the track, just wide gaps in the outside fence.

Jockey Club, Ontaio – the organization that runs the day to day operation of the track.

Knees – what are commonly called knees are the equivalent to our wrist. Eight tiny bones are arranged in two rows. Under the pounding of racing, they commonly develop arthritis.

Knock off – a quick brushing off of the horse.

Lay up – a horse that is being nursed on a farm after an injury, laid up.

Lead – in canter, there is a right and a left sequence of legs. The last foreleg to land supports the body as it reaches out further. This is the lead the horse is on.

Lead shank – a length of leather with a 30" chain with snap on one end, used to control the horse.

Legging a horse up – getting them fit with slow work.

Leg up – racehorses are not mounted from the ground. The stirrups are too high. Therefore, the rider bends his left knee and is lifted up to the saddle.

Lip chain – the mobile upper lip is lifted and the chain of the lead shank is pressed against the gums and held there by pressure put on the shank. This is a severe restraint.

Lip tattoo – every race horse's official registration number is tattooed under their top lip.

Lost one's best friend – lost one's maiden, now has to run in tougher company.

Low man on the totem pole - person with the least seniority.

Lunge line – usually a nylon strap about thirty foot long with a snap at one end.

Lunging – sending a horse around you in a circle on the end of a long line.

Maiden – non winners of a race.

Meat wagon – colloquial term for the horse ambulance.

Meet – the race days at each track.

Mud tail – tying up the length of tail to prevent it from absorbing a great deal of mud. A horse has bone in his tail about one foot from his body and the hair is wound around this bone and secured with a knot or twine.

Neck yoke – a strap around the horse's neck to hold a martingale in place.

Okayed from the gate – every horse must be broke to enter and leave the gate safely. They must show a work from the gate.

Off – slightly lame, probably not enough to actually be able to diagnose where the problem is. Irregular gait.

Off track – not fast. Muddy or drying out.

Glossary

Outdistanced – beaten by such a large margin that it is not worth calculating.

Outrider – an official who polices the track and catches runaway horses. His horse is also referred to as a pony. The head outrider leads the post parade.

Overnights – a list of the horses entered in the races in two days hence.

Packing feet with clay – to cool the feet and draw out soreness. Done after races and works.

Paddock – the area where horses are saddled for racing. To paddock a horse means to take one over to the races.

Pecking order - horses are never equal in a herd. There is always a dominant horse and they work out their relative position in status.

Pedals - stirrups.

Physic – giving a horse a remedy that 'clears his system' causing stinking diarrhea.

Place bet – betting that a horse will finish first or second. Pays less than winning.

Plate – the Queen's Plate, the most prestigious race in Canada, reserved for three year olds.

Picking feet - manure left in hooves causes a fungus called thrush. Stones can lodge in the crevasses. Either can cause lameness.

Picking stalls – removing anything obviously dirty or wet, to make stalls easier to muck the next day and keep the horse cleaner.

Pin head – apprentice jockey. A term of derision.

Polo wraps (or bandages) – stretchy bandages wrapped on horses legs for support during training.

Pony – any horse used on the track to lead another horse.

Pony Club – an organization to teach children to ride and look after their horses.

Poor doer - a horse that is hard to keep weight on. consequently, races badly.

Post parade – after the horses leave the paddock, they pass before the grandstand so the public can get a final look at them before the race.

Poultice – a clay like substance applied to the legs, then wrapped with a bandage to draw out heat and soreness.

Pulling combs – horses manes are not cut. Instead, the hair is pulled out to thin and shorten it. A pulling comb is made of metal with teeth about one inch long.

Purse - the money reward for finishing in the top four places in a race.

Racing silks - each owner has a unique coloured 'silk' (jacket and helmet cover) to identify his horses in a race.

Rate – to rate is to control the speed of the horse.

Receiving barn – a barn set aside for horses shipping in for that days racing. Empty the rest of the time.

Ringer – the practice of substituting a lookalike horse for another in a race. Usually done to cash a bet. Tattooing eliminates this practice.

Ringworm – an extremely contagious fungal skin condition. It starts as a lump on the skin which peels off in a few days, leaving a greasy circular bald spot, which continues to spread out from the centre.

Run down – a horse may stretch his legs so that his fetlock rubs on the track, causing a friction burn. It is very painful. Bandages are used to prevent the burn.

Run-in shed – a barn open on one side, with an attached paddock. The horses can take shelter when they wish but are loose in their paddock all the time.

Running a horse – taking it to the paddock for a race.

Rubbing – grooming.

GLOSSARY

Seedy toe – a condition in which an infection develops at the toe between the hoof wall and the inner structures, causing the hoof wall to separate from the foot.

Set – all the horses in a barn that go out at the same time.

Show bet – betting that a horse will finish in the top three. Pays less than win or place (second) but is a safer bet.

Shedrow – the enclosed aisle around the stalls.

Shedrowing – riding around inside the barn. Often done in bad weather.

Side reins – two straps that are fixed from the bit to the saddle, to teach the horse how to yield to the pressure of the reins.

Snaffle – a mild bit that is usually jointed in the centre.

Sock – white on the lower leg.

Spider bandage – a piece of fabric with a series of ties on both sides, usually used for knees. Its advantage is that the bandage can conform to the shape of the joint.

Splints – a tear of a ligament holding a vestigial bone to the main leg (cannon) bone and the resulting calcification. Depending on the location, the lump can cause transient lameness and an unsightly bump to more severe problems. Splints almost always occur on the inside of a front leg, due to strain.

Stall man - the person who assigns stalls to the various trainers. He also organizes the loading of the transfer vans.

Stall walking – a vice in which the horse continuously walks in circles in his stall.

Stake money – ten percent of the purse goes to the jockey, ten percent goes to the trainer. Out of the trainer's portion, the groom of the winning horse traditionally receives a small portion, usually about fifty dollars or one percent of a stake race. Many trainers withheld the money until the end of the season, after most of the help had quit or been laid off.

Stake race – a race which has a fee to be entered but pays back a higher amount of money. An added money race.

Standardbred – the breed of horses that pull sulkies for racing. They trot or pace and are not ridden.

Stands – at stud.

Stick – whip, crop.

Sweat – putting a substance on a leg, wrapping with plastic and bandaging over it. This causes the legs to sweat and draws down any filling.

Tag – claiming price.

Tapping – the practice of using a needle to draw off the excess joint fluid. The joint is then injected with cortizone.

Tattoo – the permanent identification number pressed into the underside of the top lip.

Thoroughbred – the only breed of horses allowed to race at Woodbine. The stud book (record of breeding) has been 'closed' (no outside breed allowed to cross with) for a couple of hundred years.

Three eights pole – distance is computed from the finish line, thus three eights of a mile before the wire.

Throw – to pass on a characteristic.

Tie up – see tying up.

Tightener - a work just before a race to get the horse as fit as possible.

Time sheet – every horse that works in the morning has their time recorded by one of the official clockers, then all of the times are listed on a sheet, published by the office and available later that day in a holder outside the office. This ensures all horses are fit enough to race.

Transfer vans – large horse vans, holding 12-15 horses, to transport horses to the races at a different track, if the races are not on the track where the horses are stabled.

Turning left – hot walking.

Twisting an ear – a severe method of restraint.

Twitch – a loop or rope or chain on a long stick, applied to the upper lip as a restraint.

Two minute lick – a faster canter, not yet a gallop.

Tying up – a metabolic conditions in which the muscles seize up. This can cause permanent damage to the horse. It is considered a medical emergency.

Up the track - unplaced. Run badly.

Vice – a habit the horse has no control over.

Washing out – profusely sweating. This can disrupt the delicate salt balance in the body, causing dehydration, muscle cramps, etc.

Weaving – an obsessive habit in which the horse stands at his stall door and sways back and forth on his front legs. Eventually, this will cause damage to the legs.

Webbing – during the day, the racehorses are allowed the freedom to have the head and neck out of the stall to look around. Originally, the webbing that was hung across the front of the stall was made of heavy fabric. Now it is plastic.

Weight - each horse is given an assigned weight to carry in a race. If the jockey plus tack is less than the weight, lead bars are added in pockets in the saddle towel behind the saddle.

Win bet – the wager only pays off if the horse places first.

Wire – finish line.

Withers – the bony raised area at the base of the neck. Because they are formed of elongated spinal projections, they must not have saddle pressure on them.

Working a horse – sending them full out for a certain distance. The time is recorded by the clockers.

Work sheet – a daily published sheet showing every horse that worked, the distance and time recorded.

Worming – all horses have intestinal parasites (worms) and are usually treated three or four times per year.

Wrong way – horses gallop clockwise on the track. Horses that are working stay near the inside rail. Horses that are walking home or jogging are on the outside rail, walking toward the galloping horses.

Yearlings – young horses, just one year old by the calendar, are started under tack so as to be ready to race at two.

Young rider – a division for showing horses with Olympic aspirations for riders age 16-21.

Appendix 1 - Format of Racing

Thoroughbred racing in a business that is almost 350 years old and is in a constant state of flux. At the time of this story, it was run by a combination of The Ontario Jockey Club, the Ontario Racing Commission and The Horseman's Benevolent and Protective Society.

The Jockey Club was a non-profit corporation that actually ran the racing. They owned the tracks but had to apply to the government for the dates of racing. The Jockey Club owned three Thoroughbred tracks in Ontario, Woodbine, Greenwood and Fort Erie. While the meet at Fort Erie from May to July was a 'B' meet (lower valued horses and lowered value purses), most of the racing was 'A' meet. The season started in March at Greenwood, moved to Woodbine for May to July, down to Fort Erie in August, back to Woodbine from September and October and finished back at Greenwood in November and early December.

The Jockey Club also keeps the registry of all Thoroughbreds bred and their racing records.

There were a number of officials that worked for The Jockey Club. The most important were the stewards. There were three of them. They ensured that the rules were followed and made the decisions on racing, whether infractions had occurred and what the penalty would be. One always oversaw the draw for the races, ensuring its honesty and that the star system (horses that had been entered but not drawn in) was enforced. They disciplined jockeys for riding infractions (being overweight or causing unsafe riding, for example), owners that had not paid bills and any other employees who broke the rules. The stewards were the ultimate arbitrators for the Jockey Club. Their word was law.

The clerk of the scales was responsible for ensuring all horses carried the correct weight while racing. He also oversaw the draw if there were competing claims for a horse. The shake was carried out fifteen minutes before post time. Coloured peas were assigned to each claim, put in a bottle and the winner was shaken out of the bottle.

The horse identifier was the person in the paddock who checked each horse's tattoo against his registration papers to ensure the horse was the correct one.

The valets were employed by Woodbine and assigned to the jockeys. They carried the saddles plus weights and saddle towels to the trainers and assisted in the saddling.

The racing secretary was the person who wrote up the condition book and conducted the draw for the races. **Conditions** include the date of the race, type of race (stakes, allowance, claiming), price running for if claiming, age, sex, weights carried, distance to be run, surface (dirt or turf), restrictions (ie Canadian bred) and allowance lines (non-winners since a certain date). Using predetermined guidelines, he determines how many horses may run in any race. For the trainers, the racing secretary became the most important person on the backstretch.

The Ontario Racing Commission was the branch of the provincial government that oversaw racing, keeping it honest for the public's benefit. They licenced all the owners, employees (doing criminal background checks on them), oversaw the betting, had their own veterinarian to assess the soundness and condition of the racehorses and drug tested the racehorses.

The Horseman's Benevolent and Protective Society was an association of the owners and trainers that lobbied on their behalf, acting as a sort of union. It tested the trainers for their licenses. It made sure that they had a voice at the table when changes were being made to racing, protecting their interests. It also helped out

Appendix 1 - Format of Racing

those unfortunate people in financial straits with temporary assistance.

The Canadian Thoroughbred Horse Society was formed to promote Thoroughbreds in Canada. They represented the breeders, maintained the official registry, ran the yearling sales and were involved in all aspects of Thoroughbreds, not just racing.

It takes 11 months from the time a mare is bred until she foals. The foal stays with its dam from 4-6 months before it is weaned. For the next year, the youngster continues to grow, averaging a pound or more of weight gain per day as a yearling.

Towards the late summer, the yearling sales are held. Many breeders breed just for this lucrative market. These young horses are not even broken; the purchase price is based on the perceived potential, as well as the interest of the buyers.

In the fall, most of the yearlings are started under saddle, a process which usually takes about 6 weeks. Then (at least in Ontario) they are turned out again for a few more weeks.

January 2 is the traditional date to start the racehorses back into training. While the older horses can be fitted up in a matter of weeks, the two year olds (January 1 is their 'birthday') require at least 4 months of uninterrupted training to be ready to run. Most will succumb to viruses (snotty noses) or bucked shins before they race. By now, the owner has been paying the considerable bills for 3 years before even having a chance to see his horse run.

The two year old races start at 5 furlongs, just over half a mile. A two year old at the first of the year is about equivalent with a preteen of eleven to thirteen years of age. This is why two year olds never run against older horses and if three year olds do, they are given a weight allowance. Thoroughbreds mature at 4 years of age. Many other breeds mature even later. The racing horse has been selected to mature earlier.

In Ontario at the A meet, if a horse has not broken it's maiden by the time it is four years old, it will not be given a stall and must move to another track if it is to continue racing. Most horse are off the track by the end of their three year old year. Four year olds and up were considered older horses.

The most difficult aspect of racing to understand is the claiming races. All horses running in claiming races are for sale for the amount of the claiming race. Most people wonder why anyone would run their horses 'for a tag', as we say. The point of racing is to make money. By running a horse in a claiming race, this assures lesser animals have a chance to make money as well as the owners can recoup some of their investment.

Not anyone can claim a horse. Originally, it has to be a licenced owner or trainer who has already run a horse in that meet. Later, the Jockey Club created a way for new owners to get into racing by buying a license to claim. In actual fact, only a small percentage of horses are claimed. It is fraught with inherent issues. The horse is deemed to have been bought once the gate opens and the race begins. If the horse breaks down in the race, the new owner has just bought a cripple. When a nice horse drops down in price in the claiming ranks, the dilemma becomes, is the trainer trying to steal an easy win or unload a horse breaking down? Once claimed, the horse must run for a price tag of twenty five per cent higher if he races within the first month. Horses that are not running are costing money with no chance to recoup costs.

APPENDIX 2 - WOODBINE

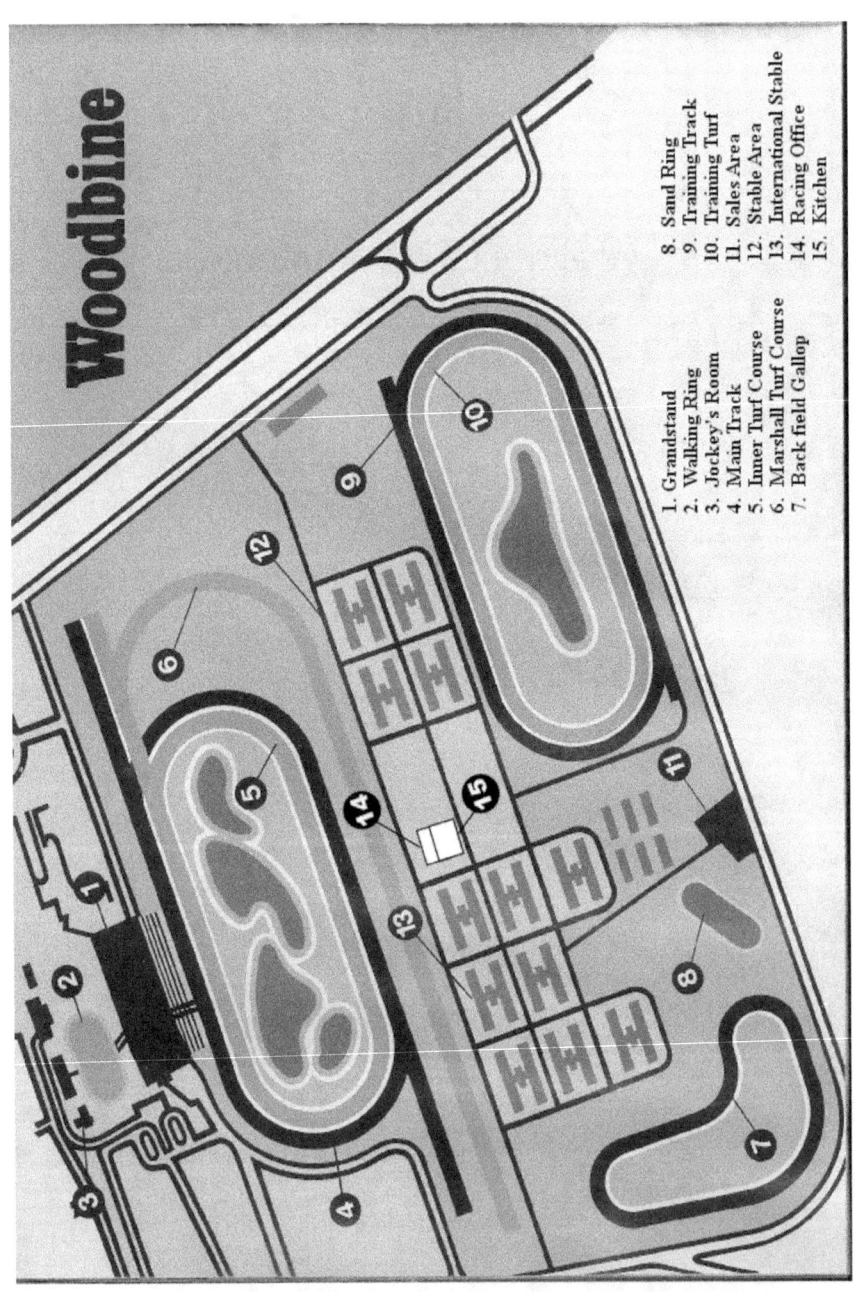

Acknowledgements

Someone asked me, "Why write your book now?" Thirty years ago, when this story was fresh, I did write some of the stories in this book. However, I did not have the resources or the connections at that time to publish a book. It takes a team to produce a book. I had to wait for technology to evolve and to make connections to find a team. Self publishing has also made books easier to bring to market. I believe the time lapse has given me a better perspective and resulted in a stronger story.

If it were not for my friend Rosanne Dolan, this book might still not have been written. She has encouraged me, reading the first draft while offering feedback and connecting me to my editor. She is my cheering section and her enthusiasm has kept me at the keyboard until the book is finished. My heartfelt thanks for your support.

I can't say enough about my brilliant editor Ramona Gorsky. The story that has ultimately been crafted is very different from the original version. Ramona has taught me how to craft a story and she has an unfailing instinct for when I am avoiding the unpleasant emotions that bring the book to life. She has balanced Rosanne's cheer-leading with pointed criticism. Ramona saw a gem in the muck and helped me find the story. I am deeply indebted to your contribution and value it highly.

My husband Ward Edwards has delighted in being responsible for the overall look of the book. He has taken charge of all the technical parts which I have no skills in doing and dread undertaking. He shows an artistic flair for the layout and has enhanced photos. I am glad to hand over this portion of the work to him. We have make a great team and I hope to draw upon his skills for future projects.

I am thrilled to have found artist Nola McConnan of Merriweather Studio for the art which graces the cover and the sketches for headers of each chapter. Few people can capture horse images realistically but I believe Nola is one of the best. Her suggestions for the artwork were so much better than my ideas and she has captured the spirit of the book completely. Because of our mutual background in the horse industry, I feel I had found one of my tribe and value all the help she has offered me.

I am lucky to find Angela Marks who has cleaned up my degraded photos as best as humanly possible. She also edited my win photos that I used by cutting out the many people that were either unknown onlookers or people whom I might have hurt by naming them in my story. My thanks to you.

Barbara Saxberg, a former client of mine, graciously went through the finished book and corrected the many little errors that had still slipped in. She also provided me with the essence of the back cover and biography that grace the cover of my book. Thanks to you, Barb, the finished product is very professional. I am also grateful to renew a friendship that had lapsed only because of the distance between us.

Former racing secretary Chris Evans spent an afternoon answering my questions on the setup of racing, especially about the claiming system and trying to make that opaque system more accessible. He also copied a contemporary map of Woodbine from my time period for me to include in the book. It is hard to reverse the lessons learned on the backstretch of keeping my nose out of other's business to now satisfy my curiosity and just ask. Thank you, Chris.

Tom Cosgrove, archivist at Woodbine, fitted me into his incredibly hectic schedule and showed me the hidden business side of Woodbine. His viewpoint helped me understand how Woodbine

ACKNOWLEDGEMENTS

has changed in the thirty years since I left. He is working hard to make thoroughbred racing relevant and fun to new generations of racegoers and I wish him the best of luck. My hope is that racing thrives into the future and more people feel that incredible thrill of watching the racehorses up close. I hope my book pays homage to your efforts for me.

I am indebted to many friends who supported me through these sometime difficult years. Sandy Battle, Cindy Conroy and Liz Ashton have given me permission to use their names. Your friendships are still valued today.

I also thank all the other friends from my track years who taught me so well.

About the Author

Janice Gannon is an accomplished horsewoman and riding instructor who has more than forty years experience in the industry. She began her lifelong love affair with horses as a child and later graduated from Humber College's School of Horsemanship. An early job in the horse industry took her to the racetrack where she dived into the little known world of the backstretch, grooming and exercising horses, working on various racetracks through Canada and the United States. She saw the best and worst of horse care, determining for herself the essence of good horsemanship. As with people, she discovered that adapting to the individual quirks and characteristics of horses was more likely to lead to a successful partnership. And succeed she did, helping to bring more than a few sour or difficult horses around to the winner's circle. Ten years later, after working with a wide variety of human and equine characters, she moved on to successfully showing her own horses, and schooling others in an empathetic approach to riding and training. After earning coaching certificates in both English and Western riding, she developed a unique style of teaching that focuses on the partnership between rider and horse. Her students value her approach of schooling horse and human together to become a smooth and happy unit. Tales from the Track is her first book.

Cover art by Nola McConnan-www.merriweatherdesignstudio.com

www.ingramcontent.com/pod-product-compliance
Lightning Source LLC
Chambersburg PA
CBHW050530300426
44113CB00012B/2026